A FIRST COURSE IN NETWORK THEORY

A First Course in Network Theory

Professor Ernesto Estrada

Professor in Mathematics and Chair in Complexity Science, University of Strathclyde, UK

Dr Philip A. Knight

Lecturer in Mathematics, University of Strathclyde, UK

OXFORD

UNIVERSITY PRESS

OXFORD
UNIVERSITY PRESS

Great Clarendon Street, Oxford, OX2 6DP,
United Kingdom

Oxford University Press is a department of the University of Oxford.
It furthers the University's objective of excellence in research, scholarship,
and education by publishing worldwide. Oxford is a registered trade mark of
Oxford University Press in the UK and in certain other countries

© Ernesto Estrada and Philip A. Knight 2015

The moral rights of the authors have been asserted

First Edition published in 2015

Published in the United States of America by Oxford University Press
198 Madison Avenue, New York, NY 10016, United States of America

British Library Cataloguing in Publication Data
Data available

Library of Congress Control Number: 2014955860

ISBN 978-0-19-872645-6 (hbk.)
ISBN 978-0-19-872646-3 (pbk.)

Printed and bound by
CPI Group (UK) Ltd, Croydon, CR0 4YY

Dedication

To Gisell, Doris, Puri, Rowan, and Finlay

Preface

The origins of this book can be traced to lecture notes we prepared for a class entitled *Introduction to Network Theory* offered by the Department of Mathematics and Statistics at the University of Strathclyde and attended by undergraduate students in the Honour courses in the department. The course has since been extended, based on experience gained in teaching the course to graduate students and postdoctoral researchers from a variety of backgrounds around the world. To mathematicians, physicists, and computer scientists at Emory University in Atlanta. To postgraduate students in biological and environmental sciences on courses sponsored by the Natural Environmental Research Council in the UK. To Masters students on intensive short courses at the African Institute of Mathematical Sciences in both South Africa and Ghana. And to mathematicians, computer scientists, physicists, and more at an International Summer School on Complex Networks in Bertinoro, Italy.

Designing courses with a common thread suitable for students with very different backgrounds represents a big challenge. For example, the balance between theory and application will vary significantly between students of mathematics and students of computer sciences. An even greater challenge is to ensure that those students with an interest in network theory, but who lack the normal quantitative backgrounds expected on a mathematics course, do not become frustrated by being overloaded by seemingly unnecessary theory. We believe in the interdisciplinary nature of the study of complex networks. The aim of this book is to approach our students in an interdisciplinary fashion and as a consequence we try to avoid a heavy mathematical bias. We have avoided a didactic 'Theorem–Proof' approach but we do not believe we have sacrificed rigour and the book is replete with examples and solved problems which will lead students through the theory as constructively as possible.

This book is written with senior undergraduate students and new graduate students in mind. The major prerequisite is elementary algebra at a level one would expect in the first year of an undergraduate science degree. To make this book accessible for students from non-quantitative subjects we explain most of the basic concepts of linear algebra needed to understand the more specific topics of network theory. This material should not be wasted on students coming from more quantitative subjects. As well as providing a reminder of familiar concepts, we expect they will encounter a number of simple results which are not typically presented in undergraduate linear algebra courses. We insist on no prerequisites in graph theory for understanding this book since we believe it contains all the necessary basic concepts in that area to allow progress in network theory. Based on our accumulated experience in teaching courses in network theory in different environments, we have also included chapters which address generic skills with which students often have difficulties. For example, we include instructions on how to manipulate and present data from simulations carried out in network theory; and how to prove analytic results in this field. Knowing how useful network theory is becoming as a tool for physicists, we have also included three chapters which draw analogies between different branches of physics and networks. Some background in physics at undergraduate level will be useful for fully appreciating these chapters but they are not necessary for understanding the rest of the book.

Every chapter of this book is written using the following common scheme: (i) the aims of the chapter are clearly stated at the beginning; (ii) a short introduction or motivation of the key topics is presented; (iii) the concepts and formulae to be used are defined with clarity; (iv) concepts are illustrated through examples, both by using small, artificial networks and also by employing real-world networks; (v) a few solved problems are given to train the student in how to approach typical problems related to the principal topics of the chapter. Predominantly we focus on simple networks—almost all of our edges will be bidirectional, unweighted, and will connect a unique pair of adjacent nodes—but we will highlight significant variations in theory and practice for a wide range of more general networks.

This book will be useful for both lecturers and researchers working in the area of complex networks. The book provides researchers with a reference of some of the most commonly used concepts in network theory, good examples of their applications in solving practical problems, and clear indications on how to analyse their results. We would also like to highlight some significant features of the book which teachers should find particularly attractive. One of the most common problems encountered by teachers is how to select appropriate illustrative exercises in the classroom. Because of the large size of many complex networks, solving problems in this field is frequently left to computers. If a student is just trained to work with a black box, they miss out on properly contextualizing theory. Frequently, this inhibits the student's ability to learn how to prove results analytically. In this book, we give solved problems which teachers can easily modify and adapt to their particular objectives. We hope that some of the examples and solved problems in this book will find their way into more conventional courses in linear algebra and graph theory, as they provide stimulating practical examples of the application of abstract concepts.

In closing, we reiterate that this book is aimed at senior undergraduate and new postgraduate students with or without quantitative backgrounds. For most of the book, the only prerequisites are a familiarity with elementary algebra and some rudiments of linear algebra. Teachers of courses in network theory, linear algebra, and graph theory—as well as researchers in these fields—should find this book attractive.

Finally, we would like to thank the colleagues and students who have helped us and inspired us to write this book. In particular, we would like to thank Mary McAuley for her patience and skill in organizing our material and Eusebio Vargas for lending us his talents to produce the high quality illustrations of networks which you will find in the book.

Ernesto Estrada
Philip A. Knight

Contents

Introduction to Network Theory

In this chapter

We start with a brief introduction outlining some of the areas where we find networks in the real-world. While our list is far from exhaustive, it highlights why they are such a fundamental topic in contemporary applied mathematics. We then take a step back and give a historical perspective of the contribution of mathematicians in graph theory, to see the origins of some of the terms and ideas we will use. Finally, we give an example to demonstrate some of the typical problems a network analyst can be expected to find answers to.

1.1 Overview of networks

One cannot ignore the networks we are part of, that surround us in every day life. There's our network of family and friends; the transport network; the telephone network; the distribution network shops use to bring us things to buy; the banking network—it does not take much effort to come up with dozens of examples. Analysis of networks particularly the huge networks that drive the global economy (directly or indirectly) is a vital science, and mathematicians have been contributing for hundreds of years.

Initially, this contribution might have been considered frivolous, and for a long time network theory was the preserve of the pure and the recreational mathematician. But more recently there have been significant theoretical and practical achievements. These lend weight to the idea that every applied mathematician should include network analysis in his or her toolkit.

1.1.1 Why are networks so ubiquitous?

One answer can be that 'being networked' is a fundamental characteristic of complex systems. If we exclude 'the science of the very large, i.e. cosmology, the study of the universe' and 'the science of the very small, the elementary particles of matter', everything remaining forms the object of study of complexity sciences, 'which includes chemistry, condensed-matter physics, materials science, and principles of engineering through geology, biology, and perhaps even psychology and the social and economic sciences'.[1]

In addition, the abstract concept of a network represents a wide variety of structures in which the entities of the complex system are represented by the nodes of the network, and the relations or interactions between these entities are captured by means of the edges of the network. Examples of some of these diverse concepts are listed below.

- **Edges representing physical links**

  Pairs of nodes can be physically connected by a tangible link, such as a cable, a road, or a vein. We include the physical network behind the internet, urban street networks, road/underground networks, water or electricity supply networks, neural and vascular networks in this category.

- **Edges representing physical interactions**

  Pairs of nodes can be considered to be connected if there is an interaction between them which is determined by a physical force, such as the interactions among protein residues, or through biological interactions such as correlated behaviour between pairs of proteins to particular stimuli in protein–protein interaction networks.

- **Edges representing 'ethereal' connections**

  Pairs of nodes may be connected by intangible connections, such as the fact that 'information' is sent from one node and it is received at another, irrespective of the 'physical' trajectory followed by this 'information' such as in the Web or in a network of airports.

- **Edges representing geographic closeness between nodes**

  Nodes can represent regions of a surface and may be connected by means of their geographic proximity, such as when we connect countries in a map, patches on a landscape connected by corridors, or cells connected to each other in tissues.

- **Edges representing mass/energy exchange**

  Pairs of nodes can be connected by relations that indicate the interchange of mass and/or energy between them, such as in reaction networks, metabolic networks, food webs, and trade networks.

- **Edges representing social connections**

  If nodes are connected by means of any kind of social tie, e.g. friendship, collaboration, or familial ties.

- **Edges representing conceptual linking**

  Pairs of nodes may be conceptually connected to each other as in dictionaries and citation networks.

[1] Cottrell, A.H. and Pettiford, D.G., *Models of Structure*. In: Structure in Science and Art. Pullman, W., Bhadeshia, H. Cambridge University Press, 2000, pp. 37–47.

- **Edges representing functional linking**

  Pairs of nodes can be connected by means of a functional relation, such as if one gene activates another; or if a brain region is functionally connected to another; or even when the work of a part of a machine activates the function of another.

You may have noticed that these concepts are not completely disjoint and it is certainly the case that we may want to interpret one network from many different points of view. Some examples of these classes of networks are illustrated in Figures 1.1–1.4.

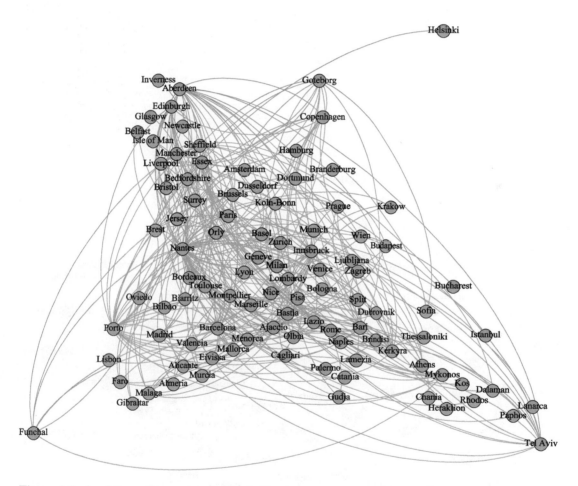

Figure 1.1 *An airline transportation network in Europe*

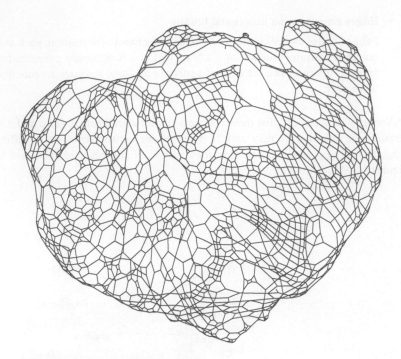

Figure 1.2 *An urban street network*

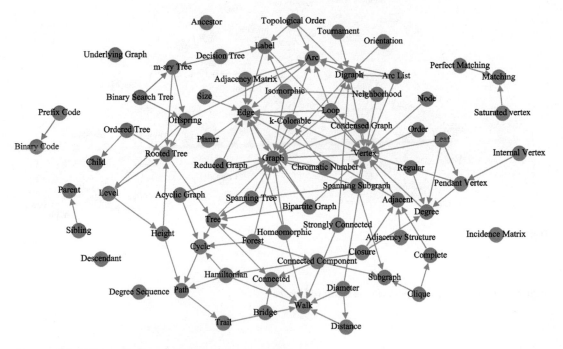

Figure 1.3 *Relational network of concepts in network theory*

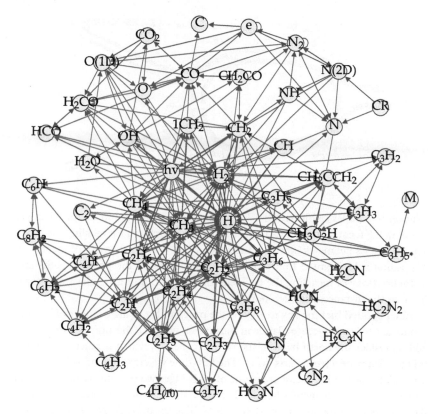

Figure 1.4 *A network of reactions in the Earth's atmosphere*

1.2   History of graphs

We present a brief history of graphs and highlight the emergence of some of the key concepts which will prove useful in our analysis of networks.

1.2.1   Euler and the Königsberg Bridge

Our story starts (as many in mathematics do) with Leonhard Euler. He had heard of a problem that the people of eighteenth century Königsberg amused themselves with.[2] The main river through the city, the Pregel, surrounds an island and branches as it flows towards the Baltic Sea. At the time, seven bridges connected the various parts of the city divided by the river. Euler's diagram of the city is shown in Figure 1.5.[3]

The problem that the city's residents tried to solve was the following: is it possible to walk through the city and cross each bridge once and only once. It is not clear how long this problem had been taxing the populace, but nobody had found a suitable path when Euler entered the picture. He described the problem and his

[2] The city is now known as Kaliningrad and is in Russia.
[3] Picture taken from Euler, L., Solutio problematis ad geometriam situs pertinentis, *Commentarii academiae scientiarum Petropolitanae*, 8:128–140, 1741.

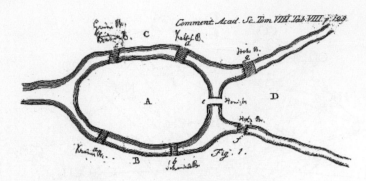

Figure 1.5 *The   seven   bridges   of Königsberg*

solution in a paper published in 1736. He was not interested in the problem *per se*, but in the fact that there was no mathematical equipment for tackling the problem despite its geometric flavour (Euler described it as an example of 'geometry of position'). He wanted to avoid exhaustively listing all the possible paths and he found a way of recasting the problem that stripped it of all irrelevant features (such as the distances between points).

Euler spotted that the key to solving the problem lay in the number of bridges connected to each piece of land; in particular, whether the number is even or odd. Consider the north bank of the river, labelled *C*, which is reached by three bridges. Suppose a path starts on *C*. The first time we cross one of its bridges we leave *C*, the second time we cross one of its bridges we re-enter *C* and the final crossing takes us out again. On the other hand, if a path does not start on *C*, after the three bridge crossings it must end on *C*. The south bank, *B*, can be reached by three bridges, and we will only finish on *B* if that is where we started.

We can conclude that each parcel of land that has an odd number of bridges must either be the start or finish of a valid route. But there are five bridges on *A*, three on *C*, and three on *D*. Since it is impossible to have a route with three end points, no valid route exists.

Euler's insight could be extended to the more general problem involving any number of bridges linking any number of pieces of land. He stated the solution as follows.

> *If there are more than two areas to which an odd number of bridges lead, then such a journey is impossible.*
>
> *If, however, the number of bridges is odd for exactly two areas, then the journey is possible if it starts in either of these areas.*
>
> *If, finally, there are no areas to which an odd number of bridges leads, then the required journey can be accomplished starting from any area.*
>
> *With these rules, the problem can always be solved.*

Euler had over-generalized in his statement: for example, suppose there is an island on the itinerary which has no bridges connected to it. Then there can never be a route including this island, no matter the number of even and odd areas elsewhere. But Euler's approach showed how network problems could be framed in such a way that they could be analysed with mathematical rigour. Implicit in his approach were the notion of a *graph*, *vertices*, *edges*, *vertex degree*, and *paths*: the atoms of network theory.

1.2.2   The knight's tour

While Euler was concerned with paths that visit every edge once and only once, the knight's tour is an example of a problem where we look for a path that visits every vertex once and only once.

If you know anything about chess, you can show very easily that it is possible to move a king around an empty chessboard so that it visits every square of the board once (using valid moves for a king in chess). The same is true for a rook and the queen; but not for a bishop (which can only visit half the squares) or a pawn, which on an empty chessboard can only move one way. But what about a knight? This is not so easy to determine, and it is quite tricky to keep track of a potential route. The problem of finding a knight's tour has been known for more than a thousand years, and solutions have been known for nearly as long (although the exact number of different tours is still unknown[4]). One such tour is given in Figure 1.6.

Euler gave a systematic treatment of the knight's tour in 1759, and this approach was extended by Alexandre-Théophile Vandermonde in an attempt to demonstrate a notation for the concept of the geometry of position. Essentially he reduced the chessboard to a set of 64 coordinates (m, n) with $1 \leq m, n \leq 8$. A knight's tour is a list of all of these coordinate pairs arranged so that if (m_1, n_1) follows (m_0, n_0) then $|m_1 - m_0| = 2$ and $|n_1 - n_0| = 1$ or $|m_1 - m_0| = 1$ and $|n_1 - n_0| = 2$.

Vandermonde showed how to exploit this notation along with some simple symmetry properties to find tours, and that by adding additional coordinates one could extend the idea to similar problems: he had in mind the possible shapes one could make by braiding threads.

The knight's tour is probably the earliest example of searching for a *circuit* in a network. A circuit that visits every vertex once and only once is known as a *Hamiltonian* circuit after the Irish mathematician William Hamilton: in the nineteenth century he had invented a board game based on finding such circuits on a dodecahedron. While it did not prove to be a spectacular success, it did popularize the notion of circuits and led to a more general analysis.

1.2.3   Trees

Network theory would have been of little interest today if it was limited to recreational applications. During the nineteenth century, though, it became apparent that network analysis could inform many areas of mathematics and science. One of the first such applications was in calculus, for which Arthur Cayley showed

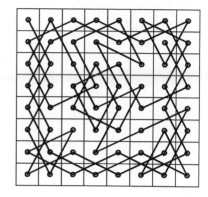

Figure 1.6 *A knight's tour*

[4] There are known to be over 26 trillion different closed tours on an 8 × 8 board which start and finish at the same square.

a connection between certain partial derivatives and *trees*. A tree is a connected network which has no circuits. We will see several in Chapter 2.

Cayley was particularly interested in *rooted trees* in which one particular vertex is designated as the root. Different choices of root can lead to different representations of the same tree. Cayley derived a method for finding the number of rooted trees with exactly n edges in terms of polynomial algebra. We omit the details but it is worth emphasizing that Cayley's algebraic approach was a key step along the way to modern methods of network analysis.

Another early analysis of trees was performed by Camille Jordan. He introduced the concept of the *centre* of a network. If one takes a tree and prunes the *leaves* (the vertices which have only one edge), one is left with another tree. Prune this tree and keep pruning and eventually you are left with a single vertex, or a tree with two vertices. These vertices can be viewed as the *centre* of the tree. Other rules for looking at trees can lead to alternative centres: we will visit the idea of *centrality* at length in Chapters 14 and 15.

1.2.4   Kirchhoff and the algebra of graphs

Matrix algebra is going to be our most useful tool for network analysis. All of the mathematicians we have mentioned so far used algebraic techniques to prove results about graphs but the methods used were introduced on a fairly *ad hoc* basis and were not easily transferrable to general problems. The person responsible for introducing matrices to the picture was Gustav Kirchhoff, a Prussian physicist (born in Königsberg).

You may have encountered Kirchhoff's laws for electrical networks. Kirchhoff formulated these laws as part of his student project when he was 21. The laws govern the way electricity flows through a network: subject to some loose constraints, the current flowing into any point must equal the flow going out and the total (directed) voltage in a closed network must sum to zero.

Kirchhoff established his law on currents by solving a set of simultaneous equations: each circuit in the network leads to an equation, but not all circuits leads to an independent equation. In the network illustrated in Figure 1.7 there are three circuits: *xy*, *yz*, and *xz* but it should be clear that there is some dependence between them: define the *sum* of two circuits as the set of the edges that belong to one but not both of them and we see that each circuit is the sum of the other two. Kirchhoff found an algebraic technique for determining and enumerating the *independent* circuits: a set of circuits no member of which is a sum of any other pair. The set of linear equations he was interested in involved this *fundamental set*.

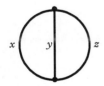

Figure 1.7 *A simple circuit*

While we can cast Kirchhoff's analysis in the language of matrix algebra, this option was not available to him—the word 'matrix' did not enter the mathematical vocabulary until the 1850s (introduced by the British mathematician James Sylvester). Kirchhoff's algebraic innovation was not appreciated by his contemporaries and it was not until the twentieth century that much of the pioneering work in network theory was revisited and recognized as an application of matrix algebra.

1.2.5   Chemistry

In this book, the terms 'network' and 'graph' are synonymous. The use of the word 'graph' in this context can be dated precisely to February 1878 where it appeared in a paper by James Sylvester entitled *Chemistry and Algebra*. By 1850, it was well known that molecules were formed from atoms; for example, that ethanol had the chemical formula C_2H_5OH. Sylvester was contributing to ensuing developments into understanding the possible arrangements of the atoms in molecules. The modern notation for representing molecules was essentially introduced by the Scottish chemist Alexander Crum Brown in 1864. He highlighted the concept of *valency*, an indication of how many bonds that each atom in the molecule must a be part of—directly related to the vertex degree.

A representation of ethanol is given in Figure 1.8. The valency of each of the atoms is clear and we can easily represent double and triple bonds by drawing multiple edges between two atoms. Notice the underlying network in this picture.

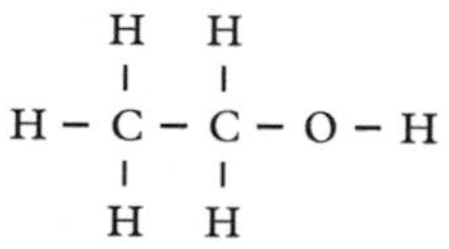

These diagrams made it plain that for some chemical formulae several different arrangements of atoms are possible: for example, propanol can be configured in two distinct ways as shown in Figure 1.9.

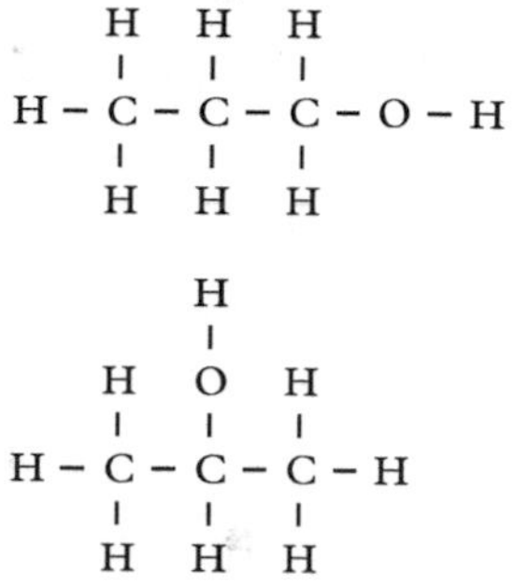

The graphical representation made it clear that such *isomers* were an important topic in molecular chemistry. Mathematicians made a significant contribution to this embryonic field. For instance, Cayley was able to exploit his earlier work on trees in enumerating the number of isomers of alkanes.[5] Each isomer has a carbon skeleton which is augmented with hydrogen atoms. Cayley realized that the number of isomers was equal to the number of different trees with n vertices (so long as none of these vertices had a degree of more than four). Sylvester built on this work by showing that even more abstract algebraic ideas could be related to the graphical structure of molecules.

The different motives of chemists and mathematicians for understanding the structures that were being uncovered meant that the disciplines diverged soon afterwards; but more recently the ties have become closer again and network theory can be used to design novel molecules with particular properties.

1.3   What you will learn from this book

In many situations in your professional life you will find problems in which network theory will be essential for their solution. In general you will be confronted with data concerning the connections between pairs of nodes of a network from which you are expected to answer specific questions. These may be about the organization of that structure; its functionality; possible mechanisms of evolution or

Figure 1.8 *Ethanol*

Figure 1.9 *Two forms of propanol*

---

[5] These are molecules with the formula C_nH_{2n+2}.

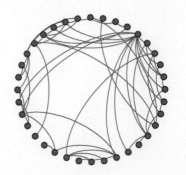

Figure 1.10 *Employee relationships in a sawmill*

growth; and potential strategies for improving its efficiency. Suppose for instance that you are analysing the network represented in Figure 1.10, which describes the social communication among a group of 36 individuals in a sawmill.

You are informed that the employees were asked to indicate the frequency with which they discussed work matters with each of their colleagues. Two nodes are connected if the corresponding individuals frequently discuss work matters. After studying this book you will be able to make the following conclusions.

1. The network forms a connected structure in which every employee discusses work matters with, on average, three or four other employees.

2. The contacts among employees do not follow a Gaussian-like distribution but instead the contacts are distributed in a rather skewed manner, with very few employees having a large number of contacts and most of the others have relatively few.

3. There is one employee (identified as Juan) who is very much central in the communication with others: Ten per cent of all the contacts in the network include him. A group of four employees, including Juan, is fundamental in passing information around the network.

4. No pair of employees is separated by more than eight steps in the network and, on average, every pair of employees is separated by only three steps.

5. The contacts among the employees are highly transitive. That is, if employee A discusses work matters with both B and C, there is a high chance (about 1/3 in this case) that B also discusses work matters with C.

6. The contacts among the employees do not resemble the pattern expected if they were created at random. Even a random model which mimics employees' preferences for limited interactions with others does not explain all the organizational complexity of this small network.

7. Employees with a higher number of contacts do not interact much between themselves. Instead, they prefer to discuss work matters with employees having very few contacts. That is, there is a certain disassortativity in the way in which employees communicate.

8. There is a central core of employees who 'dominate' the communicability in the social group.

9. The network can be divided into two almost disjoint groups of employees, in which there are preferential connections between the two groups and very few links inside each group.

10. Although the employees are Spanish-speaking (H) or English-speaking (E), which is important for communication purposes, the network is not split into such ethnic-driven communities. Instead, there are two clusters of employees, one formed by all the employees working in a section where the logs are planed (11H + 3E) and the other employees who work either in the mill, the yard, or in management (22H + 5E).

With this analysis in hand you will be able to understand how this small firm is organized; where the potential bottlenecks in the communication among the employees occur; how to improve structural organization towards an improved efficiency and functionality; and also how to develop a model that will allow you to simulate your proposed changes before they are implemented by the firm.

Enjoy the journey through network-land!

....

FURTHER READING

Barabási, A.-L., *Linked: The New Science of Networks*, Perseus Books, 2003.

Biggs, N.L., Lloyd, E.K., and Wilson, R.J., *Graph Theory* 1736–1936, Clarendon Press, 1976.

Caldarelli, C. and Catanzaro, M., *Networks: A Very Short Introduction*, Oxford University Press, 2012.

Estrada, E., *The Structure of Complex Networks. Theory and Applications*, Oxford University Press, 2011.

2 General Concepts in Network Theory

In this chapter

We introduce the formal definition of a network and some of the key terms and ideas which we will use throughout the book. We look at matrix representations of networks and make use of the adjacency matrix to identify certain network properties.

2.1 Formal definition of networks

We have seen that networks can appear in a variety of guises but they can always be thought of as a collection of items and the connections between them. In order to analyse networks, we need to turn this loose statement into formal mathematical language. To do this, we first introduce some basic set notation.

Let V be a finite set and let $E \subseteq V \otimes V$, whose elements are not necessarily all distinct. E is *reflexive* if $(v, v) \in E$ for all v in V, it is *anti-reflexive* if $(v, v) \notin E$ for all v in V and it is symmetric if $(v_1, v_2) \in E \iff (v_2, v_1) \in E$.

Definition 2.1 *A **network**, G, is a pair (V, E). Networks are also known as **graphs**. V is called the **vertex set** of G; its elements are the **vertices** of G (also known as **nodes**).*

- *If E is symmetric then G is an **undirected** network.*
- *If E is symmetric and anti-reflexive and contains no duplicate edges G then is a **simple** network.*
- *If E is nonsymmetric then G is a **directed** network (or **digraph**).*

2.2   Elementary graph theory concepts

Example 2.1

We introduce two very simple networks which we will use to illustrate some of the concepts in this chapter. They can be represented diagrammatically as in Figure 2.1.

A natural vertex set to use for both networks is $V = \{1,2,3,4\}$. Applying these vertex labels to the nodes from left to right, the edge set of G_l is then

$$E_l = \{(1,2),\,(1,3),\,(2,1),\,(2,3),\,(3,1),\,(3,2),\,(3,4),\,(4,3)\},$$

and that of G_r is

$$E_r = \{(1,1),\,(1,2),\,(1,3),\,(2,1),\,(2,3),\,(2,3),\,(2,3),\,(3,1),\,(3,2),\,(3,2),\,(3,2),\,(3,4),\,(4,3)\}.$$

Both networks are undirected. The one on the left is simple.

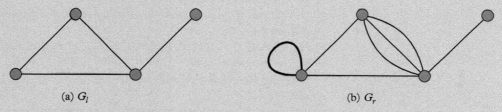

(a) G_l (b) G_r

Figure 2.1 *A simple graph (G_l) and a pseudo-multigraph (G_r)*

Neither the labelling nor the diagrammatic representation of a network is unique. For example, G_l can also be expressed as (V, E_m) where

$$E_m = \{(1,2),\,(1,3),\,(1,4),\,(2,1),\,(2,4),\,(3,1),\,(4,1),\,(4,2)\}.$$

Or as in Figure 2.2.

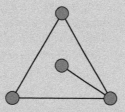

Figure 2.2 *An equivalent representation of G_l*

Sometimes when we analyse a network we are interested in the properties of individual vertices or edges—in a rail network we may be interested in the capacity of a particular station and the traffic that passes through. Sometimes, though, we are primarily interested in the overall structure of the network—how closely does the rail network resemble the road network, for example. To distinguish between the primacy of individual vertices/edges and structure as a whole we can view networks as *labelled* or *unlabelled,* although in many cases we use labels to simply be able to identify particular parts of a network.

Before proceeding we list some features of graphs that are worth naming.

Definition 2.2

- $G_s = (V_s, E_s)$ *is a* **subgraph** *of* $G = (V, E)$ *if* $V_s \subseteq V$ *and* $E_s \subseteq V_s \otimes V_s \cap E.$
- *A* **loop** *in a network is an edge of the form* (v, v). *A simple network has no loops.*
- *Suppose* $e = (v_1, v_2)$ *is an edge in the network* $G = (V, E)$. v_1 *is incident* **to** *and* v_2 *is incident* **from** *e and* v_1 *is adjacent to* v_2.

   Note that if v_1 *is adjacent to* v_2 *in an undirected network then* v_2 *is adjacent to* v_1.
- *The networks* $G_1 = (V_1, E_1)$ *and* $G_2 = (V_2, E_2)$ *are* **isomorphic** *if there is a one-to-one correspondence* $f : V_1 \to V_2$ *such that for all* $u, v \in V_1$, *the number of edges between* u *and* v *matches the number of edges between* $f(u)$ *and* $f(v)$.
- *If* $G = (V, E)$ *is a simple network then its* **complement** $\overline{G} = (V, \overline{E})$ *is the simple network with the same vertices as* G *and where* $(u, v) \in E \iff (u, v) \notin \overline{E}$.
- *Let* $G_1 = (V_1, E_1)$ *and* $G_2 = (V_2, E_2)$ *and suppose that* $V_1 \cap V_2 = \emptyset$. *The* **union** *of these two networks,* $G_1 \bigcup G_2$ *has vertex set* $V_1 \bigcup V_2$ *and edge set* $E_1 \bigcup E_2$. *This union will have (at least) two disjoint components.*

   G_1 *and* G_2 *are subgraphs of* $G_1 \bigcup G_2$.
- *Suppose that* $G = (V, E)$ *and* $F \subseteq E$. *Then* $H = (V, E - F)$ *is the subgraph we obtain by removing the edges in* F *from* G. *If* $\widehat{G} = (V, F)$ *then we write* $H = G - \widehat{G}$.
- *The* **out degree** *of a node is the number of edges that it is incident to and its* **in degree** *is the number of edges that it is incident from.*

   There is no difference between in and out degree in an undirected network: we call this common value the **degree** *of a node.*

Examples 2.2

(i) $G_l = (V, E_l)$ is a subgraph of $G_r = (V, E_r)$.
   (V, E_l) and (V, E_m) are isomorphic under the mapping $f: V \to V$ where $f(1) = 3, f(2) = 2, f(3) = 4, f(4) = 1$.
   $\overline{G_l} = (V, \{(1, 4), (4, 1), (2, 4), (4, 2)\})$, illustrated in Figure 2.3.
   Letting $F = \{(1, 1), (2, 3), (2, 3), (3, 2), (3, 2)\}$ gives $(V, E_l) = (V, E_r) - (V, F)$.

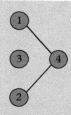

Figure 2.3 *The complement, $\overline{G_l}$, of the graph G_l*

(ii) The degrees of the vertices in G_l are 2, 2, 3, and 1. In G_r they are 3, 4, 5, and 1.

(iii) A vertex in a network which has degree 0 is referred to as an *isolated vertex*. If a vertex has degree 1 it is called an *end vertex* or a *pendant vertex*.

For example, in Figure 2.3, 3 is an isolated vertex of $\overline{G_l}$ while 1 and 2 are end vertices.

(iv) In a *complete network* (see Figure 2.4) there is an edge between every distinct node. In a complete network with n nodes, the degree of each node is $n-1$. We will use the notation K_n to denote this network.

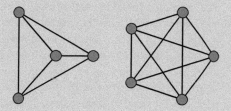

Figure 2.4 *The complete networks K_4 and K_5*

(v) A network with n nodes but no edges is called a *null graph*. Every node is isolated. We write N_n to denote this network. It is not particularly exciting, but it arises in many theorems.

(vi) A network is *regular* if all nodes have the same degree. If this common degree is k, the network is called *k-regular* or *regular of degree k*. The null graph is 0-regular and a single edge connecting two nodes is 1-regular. K_n is $(n-1)$-regular. For $0 < k < n-1$ there are several different k-regular networks with n nodes.

(vii) In a *cycle graph* with n nodes, C_n, the nodes can be ordered so that each is connected to its immediate neighbours. It is 2-regular. If we remove an edge from C_n we get the *path graph*, P_{n-1}. If we add a node to C_n and connect it to every other node we get the *wheel graph*, W_{n+1}. Examples are given in Figure 2.5.

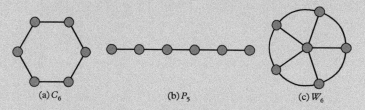

(a) C_6 (b) P_5 (c) W_6

Figure 2.5 *A cycle, a path, and a wheel with six nodes*

2.3 Networks and matrices

To record the properties of an unlabelled network, matters are often simplified if we add a generic set of labels to the nodes. For example, in a network with n vertices we can label each one with an element from the set $V = \{1, 2, \ldots, n\}$.

Definition 2.3 *Suppose $G = (V, E)$ is a simple network where $V = \{1, 2, \ldots, n\}$. For $1 \le i, j \le n$ define*

$$a_{ij} = \begin{cases} 1, & (i, j) \in E, \\ 0, & (i, j) \notin E. \end{cases}$$

Then the square matrix $A = (a_{ij})$ is called the **adjacency matrix** *of G.*

The adjacency matrix gives an unambiguous representation of any simple network. In much of the rest of this book we will look at how concepts in matrix algebra can be applied to networks and interpreting the results. Directed networks and networks with loops have adjacency matrices, too.

One or two of our examples so far have had duplicate edges. We can represent this in our adjacency matrix by letting the appropriate entries in the adjacency matrix equal the number of connections between nodes.

Any square matrix can be interpreted as a network: call a_{ij} the *weight* of an edge and the network induced by A is a *weighted network*. Weighted networks are very useful if we want to assign a hierarchy to edges in a network, but in this book their appearances are very rare.

Example 2.3

(i) Adjacency matrices for G_l and G_r (see Figure 2.1) are

$$\begin{bmatrix} 0 & 1 & 1 & 0 \\ 1 & 0 & 1 & 0 \\ 1 & 1 & 0 & 1 \\ 0 & 0 & 1 & 0 \end{bmatrix} \text{ and } \begin{bmatrix} 1 & 1 & 1 & 0 \\ 1 & 0 & 3 & 0 \\ 1 & 3 & 0 & 1 \\ 0 & 0 & 1 & 0 \end{bmatrix}.$$

(ii) If G is a simple network with adjacency matrix A then its complement, $\overline{G}$, has adjacency matrix $E - I - A$, where E is a matrix of ones.

(iii) Starting with the cycle graph C_n we can add edges so that each node is linked to its k nearest neighbours clockwise and anticlockwise. The resulting network is called a circulant network and its adjacency matrix is an example of a circulant matrix. For example, if $n = 7$ and $k = 2$ the adjacency matrix is

$$\begin{bmatrix} 0 & 1 & 1 & 0 & 0 & 1 & 1 \\ 1 & 0 & 1 & 1 & 0 & 0 & 1 \\ 1 & 1 & 0 & 1 & 1 & 0 & 0 \\ 0 & 1 & 1 & 0 & 1 & 1 & 0 \\ 0 & 0 & 1 & 1 & 0 & 1 & 1 \\ 1 & 0 & 0 & 1 & 1 & 0 & 1 \\ 1 & 1 & 0 & 0 & 1 & 1 & 0 \end{bmatrix}.$$

By taking an edge-centric view of a network, one can come up with another way of representing a network in matrix form.

Definition 2.4 *Suppose $G = (V, E)$ is a network where $V = \{1, 2, \ldots, n\}$ and $E = \{e_1, e_2, \ldots, e_m\}$ with $e_i = (u_i, v_i)$.*

For $1 \le i \le m$ and $1 \le j \le n$ define

$$b_{ij} = \begin{cases} 1, & u_i = j, \\ -1, & v_i = j, \\ 0, & \text{otherwise.} \end{cases}$$

*Then the rectangular matrix $B = (b_{ij})$ is called the **incidence matrix** of G. If e_i is a loop then we set $b_{iu_i} = 1$ and every other element of the row is left as zero.*

If the edge (u, v) is in a simple network then so is (v, u). We will only include one of each of these pairs (it doesn't matter which) in the incidence matrix. In this case, you may find in some references that the incidence matrix is defined so that all the nonzero entries are set to one and our definition of the incidence matrix is known as the **oriented incidence matrix**. There are many different conventions for including loops in incidence matrices. Since we are primarily concerned with simple networks it doesn't really matter which convention we use.

We will look more at the connections between the adjacency and incidence matrices when we look at the spectra of networks.

Example 2.4

Consider, again, G_l and G_r. G_l is simple and we can write its incidence matrix as

$$\begin{bmatrix} 1 & -1 & 0 & 0 \\ 1 & 0 & -1 & 0 \\ 0 & 1 & -1 & 0 \\ 0 & 0 & 1 & -1 \end{bmatrix}.$$

continued

Example 2.4 *continued*

For (V, E_r) the incidence matrix is

$$
\begin{bmatrix}
1 & 0 & 0 & 0 \\
1 & -1 & 0 & 0 \\
1 & 0 & -1 & 0 \\
-1 & 1 & 0 & 0 \\
0 & 1 & -1 & 0 \\
0 & 1 & -1 & 0 \\
0 & 1 & -1 & 0 \\
-1 & 0 & 1 & 0 \\
0 & -1 & 1 & 0 \\
0 & -1 & 1 & 0 \\
0 & -1 & 1 & 0 \\
0 & 0 & 1 & -1 \\
0 & 0 & -1 & 1
\end{bmatrix}.
$$

We favour matrix representations of networks in this book as they are particularly amenable to analysis and offer such a simple way of representing a network. Diagrammatic representations have a role to play—a picture can often convey information in an enlightening, evocative, or provocative way—but they can be very confusing if the network is large and they can also mislead.

2.3.1   Walks and paths

Networks highlight direct connections between nodes but indirect connections implicit in a network are frequently just as important, if not more so. If nodes are not linked by an edge, but have one (or more) nodes in common, or if there are several intermediate nodes in the way, we will usually assume that information can be passed from one to the other.

Definition 2.5 *A **walk** in a network is a series of edges (not necessarily distinct)*

$$(u_1, v_1), (u_2, v_2), \dots, (u_p, v_p),$$

*for which $v_i = u_{i+1}$ ($i = 1, 2, \dots, p-1$). If $v_p = u_1$ then the walk is **closed**.*

*A **trail** is a walk in which all the edges are distinct. A **path** is a trail in which all the u_i are distinct. A closed path is called a **cycle** or **circuit**. A graph with no cycles is called **acyclic**. A cycle of length 3 is called a **triangle**. The **walk/trail/path length** is given by p.*

We can enumerate walks using the adjacency matrix, A. The entries of A^p tell us the number of walks of length p between each pair of nodes.

Example 2.5

(i) A path of length p in a network induces a subgraph P_p and a cycle of length p induces the subgraph C_p.

(ii) In G_l we can identify a single triangle $1 \rightarrow 2 \rightarrow 3 \rightarrow 1$ (see Figure 2.1). The walk $4 \rightarrow 3 \rightarrow 2 \rightarrow 1 \rightarrow 3$ is a trail. $4 \rightarrow 3 \rightarrow 2 \rightarrow 1$ is a path. Successive powers of its adjacency matrix, A, are

$$A^2 = \begin{bmatrix} 2 & 1 & 1 & 1 \\ 1 & 2 & 1 & 1 \\ 1 & 1 & 3 & 0 \\ 1 & 1 & 0 & 1 \end{bmatrix}, A^3 = \begin{bmatrix} 2 & 3 & 4 & 1 \\ 3 & 2 & 4 & 1 \\ 4 & 4 & 2 & 3 \\ 1 & 1 & 3 & 0 \end{bmatrix}, A^4 = \begin{bmatrix} 7 & 6 & 6 & 4 \\ 6 & 7 & 6 & 4 \\ 6 & 6 & 11 & 2 \\ 4 & 4 & 2 & 3 \end{bmatrix}.$$

Note that the diagonal of A^2 gives the degree of each node. Can you explain why this is true for all simple graphs? Can you find all four walks of length three from node 1 to 3? How about all seven closed walks of length four from node 1?

Note that the shortest walk in a network between distinct nodes is also the shortest path. This can be formally established by an inductive proof.

A walk of length one must also be a path. For a longer walk between nodes u and v simply consider the walk without the last edge (that connects w and v, say). This must be the shortest such walk from u to w (why?) and therefore it is a path. This path does not contain the node v (why?) and the edge (w, v) is not a part of this path, so the walk from u to v must also be a path.

Note that the shortest closed walk is not necessarily the shortest cycle. In a simple network one can move along any edge back and forth to find a closed walk of length two from any node with positive degree. But the shortest cycle in any simple network is at least three. There are general techniques for finding the shortest cycle in any network, but we will not discuss them here.

We can find the shortest walk between any two nodes by looking at successive powers of A: it will be the value of p for which the (i,j)th element of A^p is first nonzero.

Examples 2.6

(i)   Consider C_5 (Figure 2.6). Can you write down A^2 and A^3 just by looking at the network?

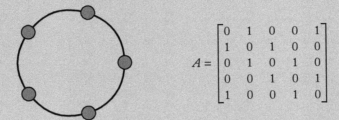

$$A = \begin{bmatrix} 0 & 1 & 0 & 0 & 1 \\ 1 & 0 & 1 & 0 & 0 \\ 0 & 1 & 0 & 1 & 0 \\ 0 & 0 & 1 & 0 & 1 \\ 1 & 0 & 0 & 1 & 0 \end{bmatrix}$$

Figure 2.6 *The cycle graph C_5 and its adjacency matrix*

(ii)   The adjacency matrix

$$A = \begin{bmatrix} 0 & 1 & 0 & 0 & 0 \\ 0 & 0 & 1 & 0 & 0 \\ 0 & 0 & 0 & 1 & 0 \\ 0 & 0 & 0 & 0 & 1 \\ 0 & 0 & 0 & 0 & 0 \end{bmatrix}$$

can be sketched as in Figure 2.7. What are the powers of A? How do you interpret this?

Figure 2.7 *A directed graph*

(iii) Consider the network in Figure 2.8.

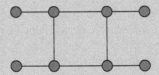

Figure 2.8 G_a, *a network with eight nodes*

Using a natural ordering of the nodes, we can write the adjacency matrix, A, and A^2, and A^3 as

$$\begin{bmatrix} 0 & 1 & 0 & 0 & 0 & 0 & 0 & 0 \\ 1 & 0 & 1 & 0 & 0 & 1 & 0 & 0 \\ 0 & 1 & 0 & 1 & 0 & 0 & 1 & 0 \\ 0 & 0 & 1 & 0 & 0 & 0 & 0 & 0 \\ 0 & 0 & 0 & 0 & 0 & 1 & 0 & 0 \\ 0 & 1 & 0 & 0 & 1 & 0 & 1 & 0 \\ 0 & 0 & 1 & 0 & 0 & 1 & 0 & 1 \\ 0 & 0 & 0 & 0 & 0 & 0 & 1 & 0 \end{bmatrix}, \begin{bmatrix} 1 & 0 & 1 & 0 & 0 & 1 & 0 & 0 \\ 0 & 3 & 0 & 1 & 1 & 0 & 2 & 0 \\ 1 & 0 & 3 & 0 & 0 & 2 & 0 & 1 \\ 0 & 1 & 0 & 1 & 0 & 0 & 1 & 0 \\ 0 & 1 & 0 & 0 & 1 & 0 & 1 & 0 \\ 1 & 0 & 2 & 0 & 0 & 3 & 0 & 1 \\ 0 & 2 & 0 & 1 & 1 & 0 & 3 & 0 \\ 0 & 0 & 1 & 0 & 0 & 1 & 0 & 1 \end{bmatrix}, \begin{bmatrix} 0 & 3 & 0 & 1 & 1 & 0 & 2 & 0 \\ 3 & 0 & 6 & 0 & 0 & 6 & 0 & 2 \\ 0 & 6 & 0 & 3 & 2 & 0 & 6 & 0 \\ 1 & 0 & 3 & 0 & 0 & 2 & 0 & 1 \\ 1 & 0 & 2 & 0 & 0 & 3 & 0 & 1 \\ 0 & 6 & 0 & 2 & 3 & 0 & 6 & 0 \\ 2 & 0 & 6 & 0 & 0 & 6 & 0 & 3 \\ 0 & 2 & 0 & 1 & 1 & 0 & 3 & 0 \end{bmatrix}.$$

Comparing these matrices, we see that the only entries that are always zero are $(1,8)$, $(4,5)$, and their symmetric counterparts. These are the only ones which cannot be connected with paths of length three or fewer. We also see, for example, that the minimum path lengths from node two to the other nodes are 1, 1, 2, 2, 1, 2, and 3, respectively.

All of this information is readily gleaned from the diagrammatic form of the network, but the adjacency matrix encodes all this information in a way that can be manipulated algebraically.

Notice that every entry of the diagonal of A^2 is nonzero. None of these entries represents a circuit: they are 'out and back again' walks along edges. Since the diagonal of A^3 is zero, there are no triangles in the network: the only circuits are permutations of $2 \to 3 \to 7 \to 6 \to 2$.

2.4   Network connectivity

A network is *connected* if there is a path connecting any two nodes. Clearly, connectivity is a significant property of a network. In an undirected network it is simple to identify connectivity: if it is not connected then there will be a part of the network that is completely separated from the rest. For a directed network it can take a little more work to confirm connectivity. At first glance it may not be obvious that there are culs-de-sac from which one cannot escape once entered.

There are a number of ways of confirming connectivity. We show how this can be done using the adjacency matrix. To do this, we will make use of *permutation matrices*: $P \in \mathbb{R}^{n \times n}$ is a permutation matrix if premultiplying/postmultiplying a matrix by P simply permutes its rows and columns. To form a permutation matrix we simple permute the rows of the identity matrix.

Problem 2.1
Show that a network with adjacency matrix A is disconnected if and only if there is a permutation matrix P such that

$$A = P \begin{bmatrix} X & Y \\ O & Z \end{bmatrix} P^T.$$

First we establish sufficiency. If A has this form then, since X and Z are square,

$$A^2 = P \begin{bmatrix} X^2 & XY + YZ \\ O & Z^2 \end{bmatrix} P^T.$$

That is, the zero block remains and a simple induction confirms its presence in all powers of A. This means that there is no path of any length between nodes corresponding to the rows of the zero block and nodes corresponding to its columns, so by definition the network is not connected.

For necessity, we consider a disconnected network. We identify two nodes for which there is no path from the first to the second and label them n and 1. If there are any other nodes which cannot be reached from n then label these as nodes $2, 3, \ldots, r$. The remaining nodes, $r + 1, \ldots, n - 1$ are all accessible from n. There can be no path from a node in this second set to the first. For if such a path existed (say between i and j) then there would be a path from n to j via i.

If there is no path between two nodes then they are certainly not adjacent. Hence $a_{ij} = 0$ if $r < i \leq n$ and $1 \leq j \leq r$, thus

$$A = \begin{bmatrix} X & Y \\ O & Z \end{bmatrix}$$

where $X \in \mathbb{R}^{r \times r}$ and $Z \in \mathbb{R}^{(n-r) \times (n-r)}$ are square.

If a network is not connected then it can be divided into *components* each of which is connected. In an undirected network the components are disjointed. In a directed network you can leave one component and enter another, but not go back. A component in a directed network that you cannot exit is referred to as *strongly connected*.

Connectivity can be associated with the number of edges in a network, since the more edges there are the more likely one should be able to find a path between one node and another.

Problem 2.2

Show that if a simple network, G, has n nodes, m edges, and k components then

$$n - k \leq m \leq \frac{1}{2}(n-k)(n-k+1).$$

The lower bound can be established by induction on m. The result is trivial if $G = N_n$, the null graph. Suppose G has m edges. If one removes a single edge then the new network has n nodes, $m-1$ edges, and K components where $K = k$ or $K = k+1$. By the inductive hypothesis, $n - K \leq m - 1$ and so $n - k \leq m$.

For the upper bound, we note that if a network with n nodes and k components has the greatest possible number of edges then every one of its components is a complete graph. We leave it to the reader to show that we attain the maximum edge number if $k - 1$ of the components are isolated vertices. The number of edges in K_{n-k+1} is $(n-k)(n-k+1)/2$.

Note that we can conclude that any simple network with n nodes and at least $(n-1)(n-2)/2 + 1$ edges is connected.

Examples 2.7

(i) Of all networks with n nodes, the complete graph, K_n, has most edges. There are $n-1$ edges emerging out of each of the n nodes. Each of these edges is shared by two nodes. Thus the total number of edges is $n(n-1)/2$. K_n has a single component.

(ii) Of all networks with n nodes, the null graph, N_n, has most components, namely n. N_n has no edges.

(iii) Consider the network $G_l \bigcup G_a$ (where G_l and G_a were defined in Examples 2.1 and 2.6, respectively). It has $n = 12$ nodes, $m = 12$ edges, and $k = 2$ components. Clearly $n - k < m$ and $m < (n-k)(n-k+1)/2$.

In $\overline{G_l}$, $n = 4$, $k = 2$ and $m = 2$ giving $n - k = m$ and $m < (n-k)(n-k+1)/2$.

In G_l, $n = 4$ and $m = 4$. Since $m > (n-1)(n-2)/2$ we know it must be connected without any additional information.

Consider a network representing social relationships. One would expect certain parts of the network to be more connected than others. That is, one would expect groups of mutual acquaintances to be linked together by more tenuous connections. This is definitely a matter that is worth analysing, so let us formalize some ideas about connectedness. We can compare networks by having a measure of connectivity. Note that this measure can be applied to each subgraph of a network.

Definition 2.6 *Suppose $G = (V, E)$ is a (simple) network.*

- *A **disconnecting set** of G is any subset $S \subseteq E$ for which the network $(V, E - S)$ has more components than G.*
- *S is a **cut-set** of G if it contains no proper subsets that are disconnecting sets.*
- *A disconnecting set is a **bridge** if it contains only one edge.*
- *The **edge connectivity** of a network $G = (V, E)$ is the number of edges in its smallest cut-set and is denoted by $\lambda(G)$. If $\lambda(G) \geq k$ then G is k-**edge connected**.*

Examples 2.8

(i) In G_a, a disconnecting set must contain an edge incident to an end vertex or two edges from the central cycle.

    A cut-set contains either one edge or two.

(ii) In the path network P_n, every non-empty set of edges is a disconnecting set and every individual edge forms a cut-set.

(iii) G_a and G_l both have an edge connectivity of one.

(iv) For the complete network K_n to become disconnected a set of nodes, say $\{1, 2, \ldots, k\}$, must become isolated from $\{k + 1, k + 2, \ldots, n\}$. For each of the k nodes, $n - k$ links need to be severed, thus a cut-set must contain $k(n - k)$ edges.

    The smallest cut-set arises when $k = 1$. Hence $\lambda(K_n) = n - 1$. This is the maximal value edge-connectivity that can be obtained by a network with n nodes.

One can also view connectivity from the perspective of vertices. A set of nodes in a connected network is called a *separating set* if their removal (and the removal of incident edges) disconnects the graph. If the smallest such set has size k then the network is called *k-connected* and its *connectivity*, denoted $\kappa(G)$, is k. If $\kappa(G) = 1$ then a node whose removal disconnects the network is known as a *cut-vertex*.

Example 2.9

(i) The left-hand network in Figure 2.9 is 1-connected. For the middle, $\kappa = 2$. k-connectivity does not really make sense in the right-hand network, K_4. We cannot isolate any nodes without removing the whole of the rest of the network, but by convention we let $\kappa = n - 1$ for K_n.

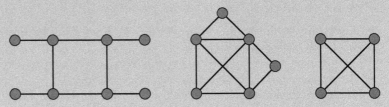

Figure 2.9 *Three networks with different connectivity properties*

Notice that for each network the minimal separating set must contain a node of maximal degree.

(ii) Figure 2.10 shows a network representing relationships between powerful families in fifteenth-century Florence. Notice that the Medici family are a cut-vertex but no other family can cut out anything other than end vertices.

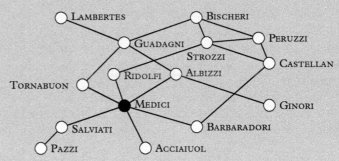

Figure 2.10 *Socio-economic ties between fifteenth-century Florentine families*

(iii) In Figure 2.11, $\kappa(G) = 2$. Can you identify all the separating sets of size two?

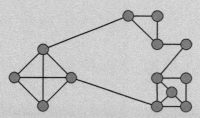

Figure 2.11 *A network with connectivity of two*

The network also has edge-connectivity of two.

(iv) If a connected network has an end vertex then $\kappa = 1$ and any node adjacent to an end vertex is a cut vertex.

Notice that in Figure 2.11 the network has a number of distinct parts which are highly connected: there are a number of subgraphs that are completely connected. This is a sort of structure that arises in many practical applications and is worth naming. Any subgraph of a simple network that is completely connected is called a *clique*. The biggest such clique is the *maximal clique*. As with many of the concepts we have seen, we can guarantee the existence of cliques of a certain size in networks with sufficient edges. It was established over a century ago that a simple graph with n nodes and more than $n^2/4$ edges must contain a triangle. This result was extended in 1941 by the Hungarian mathematician Turán who showed that if a simple network with n nodes has m edges then there will be a clique of at least size k if

$$ m > \frac{n^2(k-2)}{2(k-1)}. $$

Examples 2.10

(i) To prove his result, Turán devised a way of constructing the network with as many edges as possible with n nodes and maximal clique size k.

To do this, one divides the nodes into k subsets with sizes as equal as possible. Nodes are connected by an edge if and only if they belong to different subsets. We use $T_{n,k}$ to denote this network. $T_{5,3}$ is illustrated in Figure 2.12 for which $m = 8$.

Can you find the cliques of size four that are created by adding an extra edge?

continued

Examples 2.10 *continued*

Figure 2.12 $T_{5,3}$

(ii) $T_{n,2}$ is the densest graph without any triangles. $T_{6,2}$ is illustrated in Figure 2.13 along with the adjacency matrix, A. $T_{6,2}$ has a special structure which we will discuss in more detail.

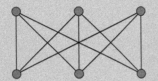

$$A = \begin{bmatrix} 0 & 0 & 0 & 1 & 1 & 1 \\ 0 & 0 & 0 & 1 & 1 & 1 \\ 0 & 0 & 0 & 1 & 1 & 1 \\ 1 & 1 & 1 & 0 & 0 & 0 \\ 1 & 1 & 1 & 0 & 0 & 0 \\ 1 & 1 & 1 & 0 & 0 & 0 \end{bmatrix}$$

Figure 2.13 $T_{6,2}$

Triangles can be identified by looking at the diagonal of the cube of the adjacency matrix. It is easy to show that $A^3 = 9A$. The zero diagonal of A^3 confirms the absence of triangles.

2.5 Graph structures

2.5.1 Trees

The word 'tree' evokes a similar picture for most people, and we can use it to describe a particular structure in a network that frequently arises in practice. We encountered trees in our history lesson.

Definition 2.7 *A **tree** is a connected network with no cycles. A **forest** is a union of trees.*

There are lots and lots of trees! There are n^{n-2} distinct (up to isomorphism) labelled trees with n nodes. For $n = 1, 2, 3, 4, 5, 6$ this gives $1, 1, 3, 16, 125, 1296$ trees before truly explosive growth sets in. Counting unlabelled trees is much harder, and there is no known formula in terms of the number of nodes but their abundance appears to grow exponentially in n.

Examples 2.11

(i) Figure 2.14 illustrates a number of trees.

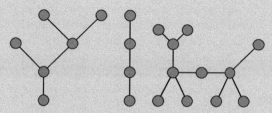

Figure 2.14 *Examples of trees*

(ii) There are only two different unlabelled trees with four nodes, as illustrated in Figure 2.15. The left-hand tree can be labelled in four ways, but only in 12 distinct ways since one half are just the reverse of the other. Once we label the pivotal node of the right-hand tree (for which we have four choices) all labellings are equivalent.

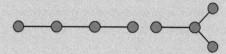

Figure 2.15 *Two unlabelled trees*

Suppose G is a connected network with cycles. Then we can break a cycle by removing an edge. The network is still connected. We can keep doing this until all cycles are removed and we end up with a tree that connects all the nodes of G. Such a tree is called a *spanning tree*.

Examples 2.12

(i) In Figure 2.16 we illustrate a network and two of the possible spanning trees.

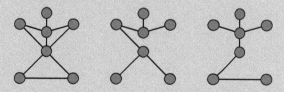

Figure 2.16 *A network and two of its spanning trees*

(ii) The notion of a spanning tree can be expanded to disconnected networks. If we form a spanning tree for each component and take their union, the result is a *spanning forest*.

(iii) The number of edges in a spanning tree/forest of G is called its *cut-set rank*, denoted by $\xi(G)$. If a network has k components $\xi(G) = n - k$. The number of edges removed from G to form the forest, $m - n + k$, is known as the *cycle rank* and is denoted by $\gamma(G)$.

For example, in (i) above, $\gamma(G) = 3$ and $\xi(G) = 6$.

2.5.2   Bipartite graphs

In any big city on a Saturday afternoon, thousands of people are out in the shops. Imagine you construct a network to record activity of business between 2pm and 3pm. Nodes represent people and an edge between two individuals represents the relationship 'enters into a financial transaction with'. Almost all edges in the network will represent a transaction between a shop-worker and a customer (exceptions would exist as some of the shop-workers may make a purchase themselves) but to all intents and purposes we should expect a network which contains two groups of nodes which have many edges between them and very few within a group.

Definition 2.8   *A network $G = (V, E)$ is **bipartite** if the nodes can be divided into disjoint sets $V_1 \bigcup V_2$ such that $(u, v) \in E \Rightarrow u \in V_i, v \in V_j, i \neq j$.*

There are many networks in real applications that are exactly or nearly bipartite. In chapter 18 we will look at how to measure how close to bipartite a network is in order to infer other properties. For now, we briefly discuss some of the properties an exactly bipartite network possesses.

Examples 2.13

(i) The Turán network, $T_{n,2}$, is bipartite. Recall that $T_{6,2}$ has the adjacency matrix

$$A = \begin{bmatrix} 0 & 0 & 0 & 1 & 1 & 1 \\ 0 & 0 & 0 & 1 & 1 & 1 \\ 0 & 0 & 0 & 1 & 1 & 1 \\ 1 & 1 & 1 & 0 & 0 & 0 \\ 1 & 1 & 1 & 0 & 0 & 0 \\ 1 & 1 & 1 & 0 & 0 & 0 \end{bmatrix}.$$

In general, if n is even, $T_{n,2}$ has adjacency matrix

$$A = \begin{bmatrix} O & E \\ E & O \end{bmatrix}$$

where E is an $(n/2) \times (n/2)$ matrix of ones.[1] It is straightforward to show that

$$A^{2k} = (n/2)^{2k-1} \begin{bmatrix} E & O \\ O & E \end{bmatrix}, \quad A^{2k+1} = (n/2)^{2k} \begin{bmatrix} O & E \\ E & O \end{bmatrix}$$

(ii) In the *complete bipartite graph* every node in V_1 is connected to every node in V_2. If V_1 has m nodes and V_2 has n we can denote this graph as $K_{m,n}$. The Turán networks $T_{k,2}$ are complete bipartite graphs, for example $T_{6,2} = K_{3,3}$.

---

[1] If n is odd, the structure is similar: $A = \begin{bmatrix} O & E^T \\ E & O \end{bmatrix}$ where E is $\frac{n-1}{2} \times \frac{n+1}{2}$.

(iii)   The *k-cube*, Q_k, is a network representing the connections between vertices in a *k*-dimensional cube. A diagrammatic representation of Q_3 is shown in Figure 2.17.

Figure 2.17 *The 3-cube Q_3*

The vertices of the unit *k*-cube have coordinates $(x_1, x_2, \ldots, x_k)$ where $x_i = 0$ or 1. Vertices are adjacent if their coordinates differ in only one place. We can divide the coordinates into those whose sum is even and those whose sum is odd. Vertices in these sets cannot be adjacent and so the network is bipartite.

The adjacency matrix, A, of a bipartite network has a characteristic structure. The division of the nodes into two groups means that there must be a permutation P such that

$$A = P \begin{bmatrix} O & X \\ Y & O \end{bmatrix} P^T.$$

In an undirected network, $Y = X^T$. It is straightforward to show that

$$A^{2k} = P \begin{bmatrix} (XY)^k & O \\ O & (YX)^k \end{bmatrix} P^T \text{ and } A^{2k+1} = P \begin{bmatrix} O & (XY)^k X \\ (YX)^k Y & O \end{bmatrix} P^T.$$

The odd powers have a zero diagonal hence every cycle in a bipartite network has even length.

Bipartivity can be generalized to *k-partivity*. A network is *k-partite* if its nodes can be partitioned into *k* sets $V_1, V_2, \ldots, V_k$ such that if $u, v \in V_i$ then there is no edge between them.

Examples 2.14

(i)   Trees are bipartite. To show this, pick a node on a tree and colour it black. Then colour all its neighbours white. Colour the nodes adjacent to the white nodes black and repeat until the whole tree is coloured. This could only break down if we encounter a previously coloured node. If this were the case, we would have found a cycle in the network. The nodes can then be divided into black and white sets. We show some appropriate colourings of trees in Figure 2.18.

continued

Examples 2.14 *continued*

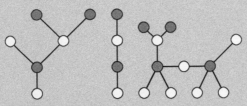

Figure 2.18 *A demonstration of bipartitivity in trees through a 2-colouring*

(ii) The maximal clique in a bipartite network has size 2, since K_n has odd cycles for $n > 2$.

(iii) The n node star graph, $S_{1,n-1}$ has a single central node connected to all other $n-1$ nodes and no other edges. $S_{1,5}$ is illustrated in Figure 2.19.

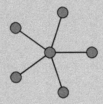

Figure 2.19 *The star graph $S_{1,5}$*

$S_{1,n-1}$ is a tree, and is also the complete bipartite graph $K_{1,n-1}$.

(iv) If G is 3-partite then its adjacency matrix can be permuted into the form

$$\begin{bmatrix} O & X_1 & X_2 \\ Y_1 & O & X_3 \\ Y_2 & Y_3 & O \end{bmatrix}.$$

Odd cycle lengths are possible.

· ·

FURTHER READING

Aldous, J.M. and Wilson, R.J., *Graphs and Applications: An Introductory Approach*, Springer, 2003.

Bapat, R.B., *Graphs and Matrices*, Springer, 2011.

Chartrand, G. and Zhang, P., *A First Course in Graph Theory*, Dover, 2012.

Wilson, R.J., *Introduction to Graph Theory*, Prentice Hall, 2010.

How To Prove It

3

In this chapter

We motivate the necessity for rigorous proofs of results in network theory. Then we give some advice on how to prove results by using techniques such as induction and proof by contradiction. At the same time we encourage the student to use drawings and counterexamples and to build connections between different concepts to prove a result.

3.1 Motivation

You may have noticed that in this book we have not employed the 'Theorem–Proof' structure familiar to many textbooks in mathematics, and when you read the title of this chapter maybe you thought, "Why should I care about proving things rigorously?". Let us start by considering a practical problem. Suppose you are interested in constructing certain networks displaying the maximal possible heterogeneity in their degrees. You figure out that a network in which every node has a different degree will do it and you try by trial-and-error to construct such a network. However, every time you attempt to draw such a network you end up stymied by the fact that there is always at least one pair of nodes which share the same degree. You try as hard as possible and may have even used a computer to help you in generating such networks, but you have always failed. You then make the following conjecture.

In any simple network with $n \geq 2$ nodes, there should be at least two nodes which have exactly the same degree.

But, are you sure about this? Is it not possible that you are missing something and such a dreamed network can be constructed? The only way to be sure about this statement is by means of a rigorous proof that it is true. Such a proof is just a deductive argument that such a statement is true. Indeed it has been said that:

Proofs are to the mathematician what experimental procedures are to the experimental scientist: in studying them one learns of new ideas, new concepts, new strategic-devices which can be assimilated for one's own research and be further developed.[1]

[1] Rav, Y., Why do we prove theorems? Philosophia Mathematica **15** (1999) 291–320.

There are many ways to establish the veracity of your conjecture. Here is a very concise argument. First observe that in your network of n nodes, no node can have degree bigger than $n - 1$. So for a completely heterogeneous set of degrees you may assume that the degrees of the n nodes of your imagined network are $0, 1, 2, \ldots, n-2, n-1$. The node with degree $n-1$ must be connected to all the other nodes in the network. However, there is a node with degree 0, which contradicts the previous statement. Consequently, you have *proved by contradiction* that the statement is true. It has become a theorem and you can now be absolutely convinced that such a dreamed network cannot exist.

When you read these statements and their proofs in any textbook they usually look so beautiful, short, and insightful that your first impression is: "I will never be able to construct something like that". However, such statement of a theorem is usually the result of a long process in which hands have possibly got dirty on the way; some not so beautiful, short, and insightful sketches of the proof were advanced and then distilled until the last proof was produced. You, too, should be able to produce such beautiful and condensed results if you train yourself and know a few general rules and tricks. You can even create an algorithmic scheme to generate such proofs. Something of the following sort.

1. Read carefully the statement and determine what the problem is asking you to do.

2. Determine which information is provided and which assumptions are made.

3. Try getting your hands dirty with some calculations to see yourself how the problem looks in practice.

4. Plan a strategy for attacking the problem and select some of the many techniques available to prove a theorem.

5. Sketch the proof.

6. Check yourself if your solution is the one asked for by the problem stated.

7. Simplify the proof as much as possible by eliminating all the superfluous statements, assumptions, and calculations.

Some of these techniques for proving results in network theory are provided in this chapter as a guide to students for solving their own problems. Have a look through Chapter 2 and you will see that we have used some of these techniques in the examples and problems. Hopefully, with practice, you can use them to solve more general problems that you find during your independent work.

3.2 Draw pictures

A well-known adage says that 'A picture is worth a thousand words' and network theory is a discipline where this is particularly true. But even though network theory is very pictorial (and this book is filled with figures of networks), it is still as analytic and rigorous as any branch of mathematics. There are many situations in which it is not obvious as to how one starts solving a particular problem. A drawing or a sketch can help trigger an idea that eventually leads to the solution. Let us illustrate this situation with a particular example.

Example 3.1

Suppose you have been asked the following.

Show that if there is a walk from the node v_1 to v_2, such that $v_1 \neq v_2$, then there is a path from v_1 to v_2.

1. You can start by sketching a walk as illustrated in Figure 3.1.

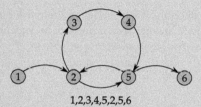

1,2,3,4,5,2,5,6

Figure 3.1 *A walk of length seven in a network between nodes 1 and 6*

2. Notice that to establish a path between 1 and 6 exists we simply have to avoid visiting the same vertex twice. For instance, avoiding visiting vertex 5 twice, we obtain the path shown in Figure 3.2(a) and if we avoid visiting node 2 twice we obtain the path illustrated in Figure 3.2(b).

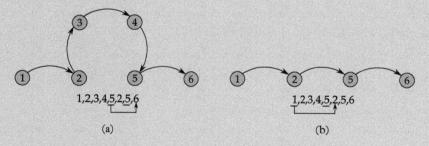

Figure 3.2 *Two walks in a network that avoid visiting a node twice*

continued

Example 3.1 *continued*

We can proceed to write these findings mathematically.

1. Let $W_1 = v_0, v_1, \ldots, v_{p-1}, v_p, \ldots, v_q, v_{q+1}, \ldots, v_{k-1}, v_k$ be a walk between v_0 and v_k.
2. If all the nodes in W_1 are distinct, then W_1 is a path and we are done.
3. If all the vertices are not different, select a pair of identical ones, say $v_p = v_q$, where $p < q$.
4. Write $W_2 = v_0, v_1, \ldots, v_p, v_{q+1}, \ldots, v_{k-1}, v_k$, which is a walk between v_0 and v_k and it is shorter than W_1.
5. If all the vertices in W_2 are distinct, then W_2 is the required path. Otherwise, select another repeated pair of nodes and proceed as before.

3.3   Use induction

Induction is a powerful technique for solving analytic problems and you have surely encountered it previously. In network theory, induction is one of the most powerful tools for solving problems. The idea is that you show something for small networks that can be inductively extended to all the networks you are studying to prove the result.

Example 3.2

Suppose that you have been asked to do the following.

Prove that a connected network with n nodes is a tree if and only if it has exactly m = n − 1 edges.

You can proceed in the following way to obtain the proof.

1. First, gain some insights by drawing some pictures as we have recommended before.
2. For the if ($\Rightarrow$) part of the theorem we proceed as follows.
   (a) Start by assuming that the network is a tree with n nodes.
   (b) We can easily verify that the result is true for $n = 1$, because this corresponds to a network with one node and zero edges, which is also a tree.
   (c) Suppose now that the result is true for any $k < n$.
   (d) Select a tree with n nodes and remove one edge. Because it is a tree, the result is a network with two disjoint connected components, each of which is a tree with n_1 and n_2 nodes, respectively.
   (e) Because $n_1 < n$ and $n_2 < n$, by the induction hypothesis the result is true for these two trees. Hence they have $n_1 - 1$ and $n_2 - 1$ edges, respectively.
   (f) As $n = n_1 + n_2$ we can verify that, returning the edge that was removed, the total number of edges in the network is $m = (n_1 - 1) + (n_2 - 1) + 1 = n_1 + n_2 - 1 = n - 1$, which proves the ($\Rightarrow$) part of the theorem.

3. For the only if ($\Leftarrow$) part of the theorem we proceed as follows.

   (a) Suppose that a connected network is not a tree and it has n nodes. Since it is not a tree it has cycles and therefore there are edges which are not bridges.

   (b) Select an edge which is not a bridge and remove it.

   (c) If the resulting network is not a tree repeat the previous step until you end up with a tree.

   (d) Because you have ended up with a tree it must have $n-1$ edges.

   (e) Because you have removed $k > 0$ edges from the network until it had $n-1$ edges you can conclude that the original network has $m > n-1$ edges and we are done.

Notice that we only used induction in the 'if' part of the proof. We are free to combine as many individual techniques as we like in creating proofs.

3.4 Try to find a counterexample

On many occasions it is good advice when trying to prove something to try finding a counterexample. It does not mean that such a counterexample exists (particularly if the result you are trying to prove is true), but in trying to build such a counterexample you find a fundamental obstruction that guides you toward the proof. In fact, many budding pure mathematicians attempting to prove their first complex result are given the following advice, which we pass on to you: spend half your time trying to prove the theorem and the other half of your time trying to disprove it. The expectation is that this holistic view of the process of proof will give you additional insight—that by trying to contradict yourself you will understand how to overcome the hurdles preventing you from a positive proof. This will work equally well if it turns out that the result you are trying to establish is false.

Example 3.3

Suppose that you have been asked to prove the following.

A network is bipartite if and only if it contains no cycle of odd length.

You can start by drawing a bipartite network (following some previous advice). Now try to add a triangle to the network. You will immediately realize that to add a triangle you necessarily need to connect two nodes which are in the same disjoint set of the bipartite network. Thus, the graph to which you have added the triangle is no longer bipartite. The same happens if you try a pentagon or a heptagon. Thus, a key ingredient in your proof should be the fact that the existence of odd cycles necessarily implies connections between nodes in the same set of the bipartition, which necessarily means destroying the bipartivity of the graph. We will see how to use this fact to prove this result using a powerful technique in Section 3.5.

3.5   Proof by contradiction

Contradiction is another powerful theoretical tool for solving problems in network theory. It is also a very intuitive and convincing method of proving a result.

Example 3.4

Let us try to prove now the result stated in Example 3.3.

A network is bipartite if and only if it contains no cycle of odd length.

For instance, assume that G is a bipartite network and that it contains an odd cycle (a contradiction according to this statement). We can then proceed as follows.

To prove necessity ($\Rightarrow$)

1. Let V_1 and V_2 be the two disjoint sets of nodes in the bipartite network.
2. Let $l = 2k + 1$ be the length of that cycle in G, such that

$$v_1, v_2, \ldots, v_{2k+1}, v_1$$

   is a sequence of consecutive nodes.
3. Assume that $v_i \in V_1$. Then, $v_{i+1} \in V_2$ because otherwise the edge (v_i, v_{i+1}) belongs to V_1, which is prohibited by the bipartivity of the network.
4. Assume that $v_1 \in V_1$, then $v_2 \in V_2$, which means that $v_3 \in V_1$, and so on until we arrive at $v_{2k+1} \in V_1$.
5. Since $v_{2k+1} \in V_1$ it follows that $v_1 \in V_2$. Thus, $v_1 \in V_1$ and $v_1 \in V_2$, which contradicts the disjointness of the sets V_1 and V_2.

We now prove sufficiency ($\Leftarrow$).

1. Suppose that the network G has no cycle of odd length and assume that the network is connected. This second assumption is not part of the statement of the theorem but splitting the network and the proof into individual components lets us focus on the important details.
2. Select an arbitrary node v_i.
3. Partition the nodes into two sets V_1 and V_2 so that any node at even distance from v_i (including v_i itself) is placed in V_1 and any node at odd distance from v_i is placed in V_2. In particular, there is no node connected to v_i in V_1.
4. Now suppose that G is not bipartite. Then there exists at least one pair of adjacent nodes that lie in the same partition V_i. Label these node v_p and v_q.
5. Suppose that the distance between v_1 and v_p is k and that between v_1 and v_q is l. Now construct a closed walk that moves from v_1 to v_p along a shortest path; then from v_p to v_q along their common edge; and finally back from v_q to v_1 along a shortest path. This walk is closed and has length $k + l + 1$, which must be odd.

6. Because every closed walk of odd length contains an odd cycle we conclude that the network has an odd cycle, which contradicts our initial assumption and so the network must be bipartite.

7. If G has more than one component then we can complete the proof by constructing sets such as V_1 and V_2 for each individual component and then combine them to form disjoint sets for the whole network. This final bit of housekeeping does not require any additional contradictions to be established.

3.6 Make connections between concepts

It is obvious advice but it is worth reminding you that network theory is an interdisciplinary field in which you can make connections among different concepts in order to prove a particular result. We actually proved the $\Rightarrow$ direction of the proof of the theorem in the last example in Chapter 2 using the algebraic connection between networks and adjacency matrices.

Example 3.5

Suppose that you have been asked to prove the following.

A network is regular if and only if $\sum_{j=1}^{n} \lambda_j^2 = n\lambda_1$, where $\lambda_1 > \lambda_2 \geq \cdots \lambda_n$ are the eigenvalues of the adjacency matrix of a network with n nodes.

Let us prove here only the statement that if the graph is regular then the previous equality holds and we leave the only if part as an exercise. We start by noticing that

$$\sum_{j=1}^{n} \lambda_j^2 = \text{tr}(A^2).$$

We also know that the diagonal entries of A^2 are equal to the number of closed walks of length two starting (and ending) at the corresponding node, which is simply the degree of the node (k_i). That is,

$$\text{tr}(A^2) = \sum_{i=1}^{n} k_i.$$

Now, we make a connection with the Handshaking Lemma, which states that $\sum_{i=1}^{n} k_i = 2m$ for any graph. Thus,

$$\sum_{j=1}^{n} \lambda_j^2 = \text{tr}(A^2) = 2m.$$

continued

Example 3.5 *continued*

We can write the average degree as: $\bar{k} = \frac{1}{n}\sum_{i=1}^{n} k_i$ and then we have

$$\bar{k} = \frac{1}{n}\sum_{j=1}^{n}\lambda_j^2.$$

Finally, because the graph is regular $\bar{k} = \lambda_1$, which proves the result.

3.7 Other general advice

There are many other techniques and tricks available for proving results in network theory and it is impossible to cover them all in this chapter. Apart from the cases we have analysed, the use of special cases and extreme examples abound in proofs in network theory. The student should learn as many of these techniques as possible to create an extensive arsenal for solving problems in this area. A recommendation in this direction is that the student tries to remember not only the theorems or statements of the results but also the techniques used in their proofs. This will allow them to make necessary connections between new problems and some of the 'classical' ones which they already know how to solve. Maybe one day one of the readers of this chapter will be able to prove a theorem that deserves the classification of a 'proof from the book'. According to Erdös, one of the founders of modern graph theory, such a book, maintained by God, would contain a perfect proof for every theorem. These proofs are so short, beautiful, and insightful that they make theorems instantaneously and obviously true. Good luck with producing such a proof!

FURTHER READING

Franklin, J. and Daoud A., *Proof in Mathematics: An Introduction*, Quakers Hill Press/Kew Books, 2011.

Hammack, R.H., *Book of Proof*, Virginia Commonwealth University, 2013.

Data Analysis and Manipulation

4

In this chapter

We turn our attention to some of the phenomena we should be aware of when carrying out experiments. For example, experimental data are prone to error from many sources and we classify some of these sources. We give a brief overview of some of the techniques that we can use to make sense of the data. For all of these techniques, mathematicians and statisticians have developed effective computational approaches and we hope our discussion gives the student a flavour of the issues which should be considered to perform an accurate and meaningful analysis. We list some of the key statistical concepts that a successful student of network analysis needs in their arsenal and give an idea of some of the software tools available.

4.1 Motivation

The focus of much of this book is to present the theory behind network analysis. Many of the networks we choose to illustrate the theory are idealized in order to accentuate the effectiveness of the analysis. Once we have developed enough theory, though, we can start applying it to real life networks—if you leaf through this book you will see examples based on complex biological, social, and transport networks, to name just three—and with a sound understanding of the theory we can ensure we can draw credible inferences when we analyse real networks. But we should also be aware of the limitations of any analysis we attempt. When we build a network to represent a food chain, can we be sure we have included all the species? When we look at a social network where edges represent friendship, can we be certain of the accuracy of all of our links? And if we study a network of transport connections, how do we accommodate routes which are seasonal or temporary?

The point is that any data collected from a real life setting are subject to error. While we may be able to control or mitigate errors, we should always ensure that the analytical techniques we use are sufficiently robust for us to be confident in our results. In this chapter, after presenting a brief taxonomy of experimental error, we look at some of the techniques available to us for processing and analysing data.

4.2 Sources of error

4.2.1 Modelling error

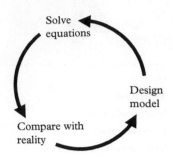

Figure 4.1 *The modelling cycle*

In almost every practical application of network theory, the network we analyse is an idealized model of a real-world situation that relies on certain assumptions holding either exactly or to within an accepted level of accuracy. We may disregard certain variables which we believe have an insignificant effect on the phenomena we are trying to understand. We may find it convenient to attach an integer value to a quantity when it would be better represented by a real number, or vice versa. Our model may be replacing a complex physical process with something which we believe is analogous, but can we be sure that the analogy holds in the circumstances where we are applying it? These are examples of some of the potential sources of modelling error. In order to control this error and build confidence in your results you should be prepared to enter the modelling cycle, illustrated in Figure 4.1. You can enter or leave the cycle at any point but it is best to go round more than once!

4.2.2 Data uncertainty

Along with the perturbations from reality introduced by the model, there are many other factors that can lead to uncertainty in the data we are working with. An obvious source of many of these errors is in the measurement of physical quantities. If the variables we are measuring can take values from a continuous set then it is inevitable that errors are introduced due to the finite accuracy of our measurements. And however carefully an experiment is carried out, some sort of underlying noise is more than likely to introduce error. These errors can usually be treated as random variables. By repeating measurements and through good experimental protocol we can often control the size of these effectively. But there are also occasions when data are missing or affected by a factor that is not random. In Section 4.3 we will introduce techniques that can mitigate against data uncertainty.

Something that is harder to deal with after data have been collected is systematic error. For example, a mistake may have been made in the units we used or a poorly thought out experimental procedure may contaminate the results. We will assume that the data we are given do not suffer from such systematic error but you should be aware of their pernicious effects when working with experimental data.

4.2.3 Computational error

Once we have our experimental data we are likely to rely on computational algorithms to perform our analysis. Numerical analysts have designed and analysed techniques for solving a vast array of mathematical problems, but you need to be aware of the limitations of the algorithms you use—and to make sure you use them appropriately. Generally, the computational techniques will give you an approximate answer to your problem. Many techniques in numerical analysis attempt to find an approximate solution by *discretizing* the problem. We split the continuum of real numbers into a discrete set of numbers separated by a step size h. We control h to achieve a balance between speed and accuracy. Whether or not you have control over the level of *discretization error* that is introduced, there is a limit on the accuracy you can expect and it is unreasonable to ask for high accuracy if the step size h is very large.

Another source of computational error is *rounding error*. On a computer, real numbers can only be assigned a finite amount of storage space, yet they may have infinite decimal expansions. We must round numbers. Each rounding error is (relatively) tiny, and usually they will have a tiny effect on the answer. Ideally, we would like the size of the perturbations introduced prior to and during computations not to be amplified significantly by our method of solution. But if the problem we are trying to solve is *badly conditioned* then the errors introduced by rounding and discretization can be amplified massively and it is possible for the accumulation of only a few rounding errors to have a catastrophic effect.

Example 4.1

Suppose we wish to compute I_{20} where $I_n = \int_0^1 x^n e^{x-1} dx$.

Note that $I_0 = \int_0^1 e^{x-1} dx = 1 - e^{-1} = 0.6321$ and using integration by parts we find

$$I_{n+1} = \left[x^{n+1} e^{x-1} \right]_0^1 (n+1) \int_0^1 x^n e^{x-1} dx = 1 - (n+1)I_n.$$

Notice that for $0 \leq x \leq 1$, if $m > n$, $0 \leq x^m \leq x^n$ and the sequence $\{I_n\}$ should be nonnegative and monotone decreasing.

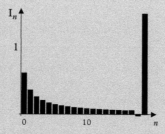

Figure 4.2 *Estimating I_n*

Figure 4.2 shows I_n as computed on a computer with 16 digits of accuracy using the recurrence $I_n = 1 - n I_{n-1}$ up to $n = 19$. At the next step it gives the value $I_{20} = -30.192$.

The problem is caused by the accumulation of tiny rounding errors (by $n = 20$ the initial error in rounding I_0 has been multiplied by 20!). The problem can be overcome by rearranging the recurrence. For example, if we let $I_n = (1 - I_{n+1})/(n+1)$ and assume $I_{30} = 0$, we calculate I_{20} to full accuracy.

4.3 Processing data

Given that the data we generate from experiments may be prone to uncertainty, it is worth considering whether we can do anything to mitigate against it. The answer will depend on the source and type of error. We now present a non-exhaustive introduction to some of the things we can attempt to do.

4.3.1 Dealing with missing data

There are multiple reasons why our observations from an experiment may be missing data. If our data come from a survey or census we are reliant on the respondents and data collectors giving complete answers. When taking measurements from a biological network it may be physically impossible to record a complete set of interactions. If data come from a series of experiments, illness or some other random misfortune may render the series incomplete. Data can be simply lost or mistranscribed. Or there may be more malicious reasons for data going missing—they may have been stolen or compromised before reaching the analyst. Whatever the reason for data going missing (and assuming we know it has) we have to make a choice—can the missing data be ignored or should it be replaced? The answer to this question, and the procedure we adopt if we plan to replace the data, will depend on whether the loss of data can be treated as a random event or not. In many cases, the simplest option of ignoring all missing data can work. But it can, of course, weaken the confidence we have in our results and may also introduce a bias, which can lead to seriously misleading conclusions.

In the context of networks, missing data essentially manifest themselves in one of two ways—missing links and missing nodes. If we cannot simply ignore the missing data, there are a number of strategies we can employ to replace it.

Example 4.2

Consider the network of sawmill employees we considered in Chapter 1. Suppose the mill is visited one year later and the study is repeated but we find that two of the employees (who we know are still working at the mill) are not recorded in our follow-up survey. To deal with the missing data we could choose one of the following options.

1. Ignore it. Maybe two missing participants will not skew our results.
2. Substitute by including the missing workers in our survey and connecting them to the network using the links to fellow employees recorded in the previous study.
3. Suppose one of the missing workers had exactly the same connections in the original survey as a worker who has not gone missing. In this case, we can substitute the missing data by assuming their links are still identical.

If there are noidentical twins (in this sense) then this process could be expanded to substitute with the links of the most similar node. Later in this book (Chapter 21) we will see ways of measuring this similarity between nodes.

4. A statistical analysis of the two networks may show some discrepancies which can be ameliorated by adding links in particular places in the network. This is the essence of a process known as *imputation*. The analysis is based on statistics such as the *degree distribution*, a concept we will also visit in Chapter 9.

For statistical reasons, imputation is the favoured approach for dealing with missing nodes as it can effectively remove bias. But dealing with the 'known unknowns' and the 'unknown unknowns' of data uncertainty can be fraught with danger for even the most experienced of practitioners and you may be playing it safe by applying a simplistic approach to missing data. For particular classes of network, sophisticated techniques exist for recovering missing data. Such techniques are beyond the scope of this book. They are underpinned by theory on high-level properties of networks such as degree distribution (see Chapter 9) and network motifs (see Chapter 13).

Example 4.3

Protein–protein interaction (PPI) networks attempt to describe affinities between proteins by measuring their tendency to interact when stimulated by a particular chemical or physical intervention. Experimental evidence of interactions is inevitably subject to noise and whether an edge (representing an interaction) should be drawn between two nodes (proteins) is not necessarily an exact science. If edges are drawn as a result of a chemical change, the simplest approach is to use a fixed threshold based on an assessment of the level of noise. However, a number of more sophisticated approaches can be developed. For example, researchers have noticed that PPI networks can be embedded in a lower dimensional geometric space than one would expect in a random network. By embedding experimental results one can thus use expert evidence to judge the likelihood of whether edges are missing.

4.3.2   Filtering data

As well as dealing with missing data, there are a number of other reasons why we may want to manipulate data before using it. For example, we may seek to eliminate repeated or rogue data; or we may want to detrend or seasonally adjust data so we can focus on particular variations of interest. Techniques for accomplishing these objectives are often referred to as filters but we reserve the term for a more restrictive class of techniques for dealing with data subject to noise (particularly unbiased noise). We define *filtering* to be the process of smoothing the data. When dealing with networks, we may want to filter raw data when determining whether to link nodes together (such as in constructing PPI networks when the results of experiments are judged) but it is also an essential tool once we start our analysis to get a smoother picture of our statistical results.

Example 4.4

In Figure 4.3(a) we show simulated data measuring the number of links (measured by total degree) within a group of individuals within a social network. The measurements are assumed to have been made indirectly and were subject to a lot of noise. The general trend appears to be that the number of links is increasing with time, but at the point indicated by the arrow there appears to be a temporary drop.

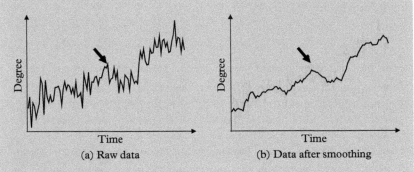

| (a) Raw data | (b) Data after smoothing |

Figure 4.3 *Applying a data filter*

To get a better idea of whether this drop is real or just an artefact of the noise we have plotted a *moving average* of the data. At every point in time we have replace the measured value with the average taken over (in this case) six successive time intervals. This 'averages out' the noise and appears to show that the apparent drop is a real phenomenon within the network.

The moving average typifies a filter where we replace a data series

$$x_0, x_1, x_2, \ldots, x_n$$

with a smoothed data set

$$y_0, y_1, y_2, \ldots, y_n$$

where

$$y_i = \sum_{j=1}^{k} a_j x_{i-k+j}. \tag{4.1}$$

There are many ways of choosing the a_j (in our example they were uniform) depending on what we are trying to achieve and the suspicions we have about the nature of the contamination of our data. For $i < k$ an appropriate filtering must be chosen to properly define the initial points in the filtered data (for example, by creating fictitious data $x_{-k}, \ldots, x_{-1}$).

4.3.3   Fitting data

Given the choice, humans naturally favour a simple explanation over a complex one. However, by assuming the role of network theorists we have to accept that complex networks yield complex results. Nevertheless, we may be able to draw links and analogies between what we see and much simpler situations. In particular, if we graph or tabulate the results of a particular network analysis it may be obvious that the data follow a particular trend, or we may expect a particular distribution of data given the analogy we have drawn. In these and many other cases we may want to fit a simple algebraic relationship between two (or more) variables. The most popular type of relationship to fit is a linear one. If this is not appropriate, we may be able to manipulate our data and change variables so that a linear fit is possible.

Example 4.5

Let x and y be two variables and a and b be constants. A simple change of variables can be used to convert some nonlinear relationships between the variables into linear ones:

$$
\begin{aligned}
y = ax^2 + b, &\Rightarrow y = aX + b, \quad X = x^2, \\
y = ae^{bx} &\Rightarrow Y = bx + \ln a, \quad Y = \ln y, \\
y = ax^b &\Rightarrow Y = bX + \ln a, \quad Y = \ln y, \quad X = \ln x.
\end{aligned}
\tag{4.2}
$$

Suppose we are given data $(x_1, y_1), \ldots, (x_n, y_n)$ for two variables. The basic principle of linear fitting is to find constants a and b so that the line $y = ax + b$ matches the data as closely as possible. We do this by minimizing the errors $e_i = y_i - ax_i - b$ over all choices of a and b. There are many ways of choosing the error measure but if we assume that the errors can be modelled by a random variable then it usually makes sense to minimize the *Euclidean distance* in the errors, namely,

$$
\min_{a,b} \sqrt{\sum_{i=1}^{n} (y_i - ax_i - b)^2},
\tag{4.3}
$$

to find the *least squares solution*, for which there are many efficient computational techniques.

If the relationship between the variables (or transformed variables) cannot be represented by a straight line then we can generalize the process and look for the best-fitting solution from a bigger class of functions (for example, polynomials of degree k, a sum of exponentials, or a trigonometric series). We can also look to fit multivariate relationships, too. Once we start our network analysis you will see that there are many occasions when many different types of relationship appear to fit the data equally well (or badly!). Whenever you attempt to find a particular fit between two variables you should ideally have some justification for it *a priori* for the results to be truly meaningful.

Having fitted a curve to data we are then in a position to *interpolate* or *extrapolate* our data. That is, we can use the curve to approximate the relationship between variables between points, or outside the range from which they have been collected. Alternatively, if we are confident in our data and do not wish to apply a fit, we can use it to interpolate or extrapolate directly. Interpolation and extrapolation can also be used to replace missing data, too.

Example 4.6

Given that the variables x and y are related, and that when $x = 0, 2$, and 3, $y = 0, 4$, and 9, respectively, we can estimate the value of y when $x = 1$ in a number of ways. For example, we can assume that between $x = 0$ and $x = 2$, the relationship between the variables is linear and hence when we interpolate to $x = 1$ we find $y = 2$. Or we notice that our three data points lie on a quadratic curve and if we assume this relationship throughout then $y = 1$ at $x = 1$. We could also fit other shapes to pass through the points or we could interpolate from the least squares linear fit to the data, $y = 0.94 + 0.93x$.

Any of our choices can be used to extrapolate outside the measured values of x but notice that these will diverge significantly as x increases.

As with all the other techniques of massaging data that we have presented in this section, we can be most confident in our interpolated/extrapolated data when there is an underlying justification for the method we use provided by the theory. If we have a large amount of data we should be judicious in the use of interpolation/extrapolation. Large amounts of data allow us to calculate unique interpolants of great complexity (for example, with 25 pieces of data we can fit a degree 24 polynomial). But these complicated functions can fluctuate wildly between the given data and give outlandish results in the gaps. Extrapolation especially needs to be done carefully. Any discrepancy between our assumed fit and the actual behaviour can be accentuated to ridiculous proportions and lead to predictions which are physically impossible.

4.4 Data statistics and random variables

Once we have analysed a large set of data we may end up producing a huge volume of results. And, of course, these results will have accumulated errors on the way. To get a feeling of what the results mean it is often useful to calculate certain representative statistics: averages, maxima, and minima. For a set of results x, some of the statistics we will make use of the most are the maximum, x_{max}, the minimum, x_{min}, the mean, $\bar{x}$, the standard deviation, σ_x, and the variance, σ_x^2. If we label the individual elements of x as $x_1, x_2, \ldots, x_n$ then

$$\bar{x} = \frac{1}{n} \sum_{i=1}^{n} x_i, \quad \text{and} \quad \sigma_x^2 = \frac{1}{n} \sum_{i=1}^{n} (x_i - \bar{x})^2. \tag{4.4}$$

It is often natural (and profitable) to look for and measure *correlations* between variables. A simple way to compare two variables x and y is to measure the *covariance* between samples $\{x_1, x_2, \ldots, x_n\}$ and $\{y_1, y_2, \ldots, y_n\}$, defined as

$$\text{cov}(x, y) = \frac{1}{n} \sum_{i=1}^{n} (x_i - \bar{x})(y_i - \bar{y}). \tag{4.5}$$

Powerful statistical techniques have been developed which boil down a large amount of information into a number between -1 and 1 which gives a precise value of the dependency between two variables. For example, if we assume that x and y are linearly related then we can calculate the *Pearson correlation coefficient* of two samples as

$$r = \frac{\text{cov}(x, y)}{\sigma_x \sigma_y}. \tag{4.6}$$

If $r = 1$ then we can infer a perfect linear relationship between x and y. If $r = -1$ then one variable goes up as the other goes down, while we can infer independence between variables if $r = 0$. In practice we would expect r to take non-integer values and the strength of the correlation should be judged against a null hypothesis. Alternatively, a whole group of samples can be compared by computing the covariances between each individual pair and forming a *covariance matrix*.

Throughout our analysis of networks we will be comparing our results from real-world networks against data generated by random variables. We are not suggesting that the real-world is random, but there is still much value in a comparison of deterministic and random phenomena. In large data sets there is often a sound mathematical reason for the distribution of the data to match that of a benchmark random variable and many of the simplified models that provide the motivation for our analysis are based on assumptions of randomness. Among the distributions we will make use of are the binomial, Poisson, Gaussian, power law, and exponential, and the student should make sure she is familiar with these. Illustrations of some of these distributions can be found in Chapter 9. In the real-world we deal with finite networks and it is natural to use discrete probability distributions with finite ranges in these models. However, for large networks it is usually safe to approximate with continuous distributions and/or ones with infinite domains such as the Gaussian. And (since we are mathematicians!) we will be taking limits to infinity to get a complete understanding of the finite. But care should be taken so that our use of random distributions does not give meaningless results and the student should be prepared to use truncated approximations to idealized distributions to avoid such misfortune.

4.5 Experimental tools

Speak to any enthusiastic mathematician for long enough and she will tell you that the subject is not a spectator sport—you need to get out there and participate to really appreciate the subject. The authors of this book share that sentiment wholeheartedly and we encourage readers to get hold of their own networks to test and develop the analytical techniques presented in this book. To that end, we give a brief presentation of some of the tools you can use for software analysis. The nature of software development means that the picture changes rapidly and this section of the book is likely to become dated very quickly. To guard against premature ageing our survey is fairly generic. We cover three general areas—sources of data, data types, and software tools.

4.5.1 Sources of data

The first step in network analysis is to get hold of a network. You can generate your own by simply creating an adjacency matrix but to get a feel for the properties of particular classes of techniques you may want to use networks which have been created and curated by previous researchers. Many of the networks we study in this book have been trawled from recent research literature and we encourage you to look at these yourself. Of course, search engines can help you find almost anything but you will find a targeted search much more productive, in particular you might want to start with online network data repositories. The dynamic nature of the web means that the data currently available could disappear at any time but typing one of the following terms (or something similar) into a search engine should be productive: 'complex network data', 'complex network resources', or 'network data repository'. In particular, at the time of going to press, we can recommend the KONECT set at the University of Koblenz which contains a rich variety of data and the SNAP set of large networks at Stanford University. You will also find that many of the authors referenced in the 'Further reading' sections at the end of each chapter have their own sets of data on their personal websites.

4.5.2 Data types

There is a lot of information in a large complex network and one needs to pay consideration as to how it should be stored to promote efficient analysis. In this regard, there is a significant difference between unlabelled and labelled networks.

Understanding unlabelled networks can give us rich insight into structure and theory but in applications the actual labels attached to the nodes must be taken into account if we want to name the most important node or the members of a particular set. For these labelled graphs, a wide array of file formats have been developed that are influenced by their author's particular interests.

For unlabelled networks one can work exclusively with an adjacency matrix. All mathematical software packages will have a format for storing matrices and an initial analysis can be performed using well-established methods from linear algebra. However, for large networks it may be necessary to use efficient storage formats. Suppose you are analysing a simple network with n nodes and m edges. To store this information one simply needs a list of the edges and their end points. If we assign each node a number between 1 and n we can store all the information as a set of m pairs of numbers. Assuming each number requires four bytes of storage (which gives us around four billion numbers to play with) the whole network therefore requires $4m$ bytes of room in the computer's memory. Unless instructed otherwise, most software packages will allocate eight bytes to store each entry of a matrix (in so-called *double precision* format). Thus the adjacency matrix requires $8n^2$ bytes of storage.

Example 4.7

Consider a social network of around 100,000 people who are each connected on average to around 100 other individuals. We only need around 20 megabytes to store the links ($m \approx 5,000,000$ since each edge adds two to the total of connections). If we form the adjacency matrix we need 8×10^{10} bytes or 80 gigabytes; a factor 4,000 times as big. While most modern computers have room for a file of 80GB, they may not be easy to manipulate (for example, as of 2014 very few personal computers would be able to store the whole matrix in RAM) and the problem is exacerbated as n increases.

The point of Example 4.7 is that even if they are simple and unlabelled networks, thought needs to go in to the method of representing a network on a computer. If we just want to store the edges then this can be done in a simple two-column text file, or a spreadsheet. It is easy enough to work in a similar way with directed networks and by adding an additional column one can also add weights.

If one makes use of a simple file type then there are usually tools to convert the network into a format which can be manipulated efficiently by your software package of choice. If the amount of storage is critical, one can consider formats that attempt to compress information as much as possible (for example graph6 and sparse6).

Simple text files can also be used to represent labelled networks but, again, when m and n are large it often pays to consider a format which is optimal for the software we want to use. Formats such as GML and Pajek have been designed to make network data portable and use flexible hierarchical structures which allow researchers to add detailed annotations to provide context, but one can also make use of some of the other countless file types that were developed without networks necessarily being at the forefront of the creators' minds.

4.5.3   Software tools

As with the formats for storing data, there is a vast array of software for analysing networks—both specialized and generic. The useful half-life of some of these packages is extremely short and so we restrict our attention to a few packages which have been around for a while.

MATLAB is a popular tool in many branches of scientific computation. Its origins are in solving problems in computational linear algebra and so it is very well suited to perform many of the operations we will discuss in this book which exploit the representation of networks in matrix form. It is not free software but is used in many universities and so is accessible to many students. MATLAB 'clones' have been developed, such as Octave, which have much of the same functionality and are licensed as free software. Since MATLAB was not designed as a tool for network analysis, it may be missing commands to compute some of the statistics you are interested in but there are active communities of MATLAB users who are constantly developing new functions and there are a number of freely available MATLAB toolboxes designed specifically for these purposes. However you acquire your data, you should be able to import it into a format that MATLAB can recognize and it has powerful graphical capabilities, albeit not necessarily tuned for illustrating complex networks.

Python users form a very active community of software developers and there are many packages that can be used to perform network analysis. In particular, most (if not all) of the network measures we introduce in this book can be calculated using the package NetworkX. Python is free software, and for mathematical computations you can use the open-source software system Sage. This can be used anywhere with an internet connection through a web browser without installing anything onto your own machine. If you have no experience of Python, you need a little bit of patience to get NetworkX up and running and doing what you want it to do. But, as the authors of this book can attest, even a fool can eventually get impressive results.

If your computing experience is limited to Microsoft Office or similar software, you can still perform sophisticated network analysis. Spreadsheets are a user friendly format for storing networks and Excel (and its rivals) have many of the mathematical and statistical functions necessary to proceed. Again, these products were not designed for working with networks, and for large sets of data, computational accuracy and efficiency may be compromised.

Finally, another open-source option is Gephi. This is designed especially for the purpose of exploring networks visually. It does not have some of the capabilities of the other packages we have mentioned for analysing very large networks but often a purely visual approach can lead one to uncover patterns which can then be analysed more systematically. In particular, it is an excellent package for creating arresting illustrations of networks and we have used it extensively in producing this book.

....

FURTHER READING

Clarke, G.M. and Cooke, D., *A Basic Course in Statistics*, Edward Arnold, 1998.

Ellenberg, J., *How Not To Be Wrong: The Hidden Maths of Everyday Life*, Allen Lane, 2014.

Lyons, L., *A Practical Guide to Data Analysis for Physical Science Students*, Cambridge University Press, 1991.

Mendenhall, W., Beaver, R.J., and Beaver, B.M., *Introduction to Probability and Statistics*, Brooks/Cole, 2012.

5 Algebraic Concepts in Network Theory

In this chapter

The primary aim of this book is to build a mathematical understanding of networks. A central tool in this is matrix algebra: there is a duality between networks and matrices which we can exploit. So before we start analysing networks, we review some results from matrix algebra and develop ideas that will be helpful to us.

5.1 Basic definitions of networks and matrices

Throughout this book we will be exploiting the fact that we can represent networks with matrices: we have already defined the adjacency matrix and the incidence matrix—and there are more to come! We can then make use of established matrix theory to deduce properties of networks. Let us first introduce some basic matrix quantities and notation which will prove useful. We will then focus on eigenvalues and eigenvectors of matrices in order to prepare ourselves for working with the spectrum of a network.

A typical matrix A will either be square (and in $\mathbb{R}^{n \times n}$) or rectangular (and in $\mathbb{R}^{m \times n}$). Its (i,j)th entry will be denoted a_{ij} and its transpose is the matrix A^T whose (i,j)th entry is a_{ji}. If $A = A^T$ then we call the matrix *symmetric*. The *trace* of a square matrix, written $\text{tr}(A)$, is defined to be the sum of the diagonal elements of a matrix.

Vectors of length n can be thought of as $n \times 1$ matrices (if they are columns) or $1 \times n$ (for row vectors). We will use bold letters to represent vectors. The inner product of two vectors $\mathbf{x}$ and $\mathbf{y}$ of the same dimension is the scalar quantity $\mathbf{x}^T \mathbf{y}$. If the inner product is zero then the vectors are said to be *orthogonal* to each other. An *orthogonal matrix* is a square matrix in which all of the columns are mutually orthogonal (and are scaled to have a Euclidean length of one). If A is orthogonal then $A^T A = I$, the identity matrix.

Occasionally we will deal with complex matrices. Most results for real matrices have direct analogues for complex matrices so long as we use the *conjugate transpose,* A^*, in place of A^T. The (i,j)th entry[1] of A^* is $\overline{a_{ji}}$. If $A = A^*$ we call the matrix *Hermitian* and if $A^*A = I$ then we call the matrix *unitary*.

We will need to make use of the determinant of some matrices. We use Laplace's definition which can be linked to salient properties of networks.

Definition 5.1 *Let $A \in \mathbb{R}^{n \times n}$ then its* **determinant,** *written* $\det(A)$ *is the quantity defined inductively by*

$$\det(A) = \begin{cases} A, & n = 1, \\ \sum_{j=1}^{n}(-1)^{i+j}a_{ij}\det(A_{ij}), & n > 1, \end{cases}$$

for any fixed i, where A_{ij} denotes the submatrix formed from A by deleting its ith row and the jth column.

The determinants of the $(n-1) \times (n-1)$ matrices used in the definition are known as *minors*. In particular, a $k \times k$ *principal minor* is the determinant of a $k \times k$ principal submatrix of A formed by taking the intersection of k rows of A with the same k columns.

The determinant has a host of theoretical uses that stem from the fact that many identities can be established, such as $\det(AB) = \det(A)\det(B)$ and $\det(A^T) = \det(A)$. The most useful theoretical property of the determinant is that it can be used to characterize singular matrices: a square matrix, A, is singular if and only if $\det(A) = 0$.

[1] Here $\overline{z}$ is the complex conjugate of z.

Examples 5.1

(i) If A is a 2×2 matrix then, letting $i = 1$ in the definition,

$$\det(A) = a_{11}a_{22} - a_{12}a_{21}.$$

And if $A \in \mathbb{R}^{3 \times 3}$,

$$\det(A) = a_{11}(a_{22}a_{33} - a_{23}a_{32}) - a_{12}(a_{21}a_{33} - a_{23}a_{31}) + a_{13}(a_{21}a_{32} - a_{22}a_{31})$$
$$= a_{11}a_{22}a_{33} + a_{21}a_{32}a_{13} + a_{31}a_{12}a_{23} - a_{13}a_{22}a_{31} - a_{23}a_{32}a_{11} - a_{33}a_{12}a_{21}.$$

(ii) Let

$$A = \begin{bmatrix} 1 & 3 & 1 \\ 2 & 1 & 1 \\ 0 & 3 & 1 \end{bmatrix}.$$

Then $\det(A) = 1 + 6 + 0 - 0 - 3 - 6 = -2$. There are three 2×2 principal minors, $\det(A_{11}) = 1 - 3 = -2$, $\det(A_{22}) = 1 - 0 = 1$, and $\det(A_{33}) = 1 - 6 = -5$.

continued

Examples 5.1 *continued*

(iii) Suppose that G is a simple connected network with three nodes. Then the adjacency matrix of G is either

$$\begin{bmatrix} 0 & 1 & 1 \\ 1 & 0 & 1 \\ 1 & 1 & 0 \end{bmatrix}$$

or a permutation of

$$\begin{bmatrix} 0 & 1 & 1 \\ 1 & 0 & 0 \\ 1 & 0 & 0 \end{bmatrix}.$$

The first matrix represents K_3, a triangle, and the second is the adjacency matrix of P_2. The determinant of the first matrix is two and that of the second is zero. In this very simple case we can characterize connected networks through their determinant.

5.2 Eigenvalues and eigenvectors

Eigenvalues and eigenvectors are the key tools in determining the action of a linear operator in many different branches of mathematics, not least in the study of networks.

Definition 5.2 *For any n dimensional square matrix A there exist scalar values[2] λ and vectors[3] $\mathbf{x}$ such that*

$$A\mathbf{x} = \lambda\mathbf{x}.$$

Any value of λ that satisfies this equation is called an eigenvalue. Any nonzero vector $\mathbf{x}$ that satisfies this equation is called an eigenvector.

For all values of λ we have $A0 = \lambda 0$. The zero vector is not considered an eigenvector, though. Now

$$A\mathbf{x} = \lambda\mathbf{x} \iff (A - \lambda I)\mathbf{x} = 0,$$

which means that the matrix $A - \lambda I$ is singular. We have therefore established the important result that the eigenvalues of λ are the roots of the equation $\det(A - \lambda I) = 0$, a polynomial of degree n.

Definition 5.3 *The polynomial*

$$\det(A - \lambda I) = b_0 + b_1\lambda + b_2\lambda^2 + \cdots + b_n\lambda^n$$

is known as the **characteristic polynomial** *(or c.p.) of A. The equation* $\det(A - \lambda I) = 0$ *is known as the* **characteristic equation**.

---

[2] At least one and at most n.
[3] At least one for each value of λ.

Once we know the eigenvalues we can factorize the c.p. so

$$\det(A - \lambda I) = (\lambda_1 - \lambda)^{p_1}(\lambda_2 - \lambda)^{p_2}\ldots(\lambda_k - \lambda)^{p_k}$$

where the λ_i are unique. Generally $k = n$ and $p_i = 1$ for all i. But sometimes we encounter *repeated eigenvalues* for which $p_i > 1$. The value of p_i is known as the *algebraic multiplicity* of λ_i. For each distinct eigenvalue there are between 1 and p_i linearly independent eigenvectors. The number of eigenvectors associated with an eigenvalue is known as its *geometric multiplicity*.

Encoded in the characteristic polynomial are certain quantities that are very useful to us. We can link its coefficients to principal minors: $(-1)^k b_{n-k}$ is the sum of all the $k \times k$ principal minors of A. Since $\text{tr}(A)$ is also the sum of the principal 1×1 minors of A, $-\text{tr}(A)$ is the coefficient of λ^{n-1} in the c.p. At the same time, if A has eigenvalues $\lambda_1, \lambda_2, \ldots, \lambda_n$ then

$$\det(\lambda I - A) = (\lambda - \lambda_1)(\lambda - \lambda_2)\cdots(\lambda - \lambda_n)$$

and multiplying out the right-hand side we immediately see that the coefficient of λ^{n-1} is

$$-\lambda_1 - \lambda_2 - \cdots - \lambda_n$$

which means that the sum of the eigenvalues of a matrix equals its trace.

For a general real matrix A, the roots of the characteristic polynomial can be complex. Notice that if A is a real matrix then

$$\det(A - \lambda I) = \det((A - \lambda I)^*) = \det(A^T - \overline{\lambda} I).$$

So if λ is a root of the c.p. of A, $\overline{\lambda}$ is a root of the c.p. of A^T. A is real though, so its complex roots appear as conjugate pairs, so $\overline{\lambda}$ must be a root of the c.p. of A, too. We can conclude from this that A and A^T have the same eigenvalues. Suppose that $A\mathbf{x} = \lambda\mathbf{x}$ and $A^T\mathbf{y} = \overline{\lambda}\mathbf{y}$, which means that $\mathbf{y}^*A = \lambda\mathbf{y}^*$ (why?). Then $\mathbf{x}$ is known as a right eigenvector of A (corresponding to λ) and $\mathbf{y}^*$ is the left eigenvector.

Problem 5.1
Show that if $\mathbf{x}$ and $\mathbf{y}^*$ are right and left eigenvectors corresponding to different eigenvalues then they are orthogonal and hence conclude that if $A \in \mathbb{R}^{n \times n}$ is symmetric then (i) the eigenvalues of A are all real and (ii) the eigenvectors of distinct eigenvalues A are mutually orthogonal.

Suppose that $\mathbf{x}_1$ is the right eigenvector corresponding to $\lambda_1 \neq 0$ and $\mathbf{y}_2^*$ is the left eigenvector corresponding to λ_2, where $\lambda_2 \neq \lambda_1$. Then,

$$\mathbf{y}_2^*\mathbf{x}_1 = \mathbf{y}_2^*\frac{A\mathbf{x}_1}{\lambda_1} = (\mathbf{y}_2^*A)\frac{\mathbf{x}_1}{\lambda_1} = \frac{\lambda_2}{\lambda_1}\mathbf{y}_2^*\mathbf{x}_1.$$

But $\lambda_2 \neq \lambda_1$ so $\mathbf{y}_2^*\mathbf{x}_1 = 0$. If $\lambda_1 = 0$,

$$\mathbf{y}_2^*\mathbf{x}_1 = \mathbf{y}_2^*\frac{A\mathbf{x}_1}{\lambda_2} = \lambda_1\mathbf{y}_2^*\mathbf{x}_1 = 0.$$

If $A = A^T$ then by symmetry $\mathbf{x} = \mathbf{y}$ in the above argument, hence the left and right eigenvectors are identical.

(i) Since A is symmetric, $(\mathbf{x}^*A\mathbf{x})^* = \mathbf{x}^*A^*\mathbf{x} = \mathbf{x}^*A\mathbf{x}$, for any vector $\mathbf{x}$. Now $(\mathbf{x}^*A\mathbf{x})^*$ is the complex conjugate of $\mathbf{x}^*A\mathbf{x}$, and for a number to equal its complex conjugate it must be real. Now suppose that λ is an eigenvalue of A with eigenvector $\mathbf{x}$. Then $\mathbf{x}^*A\mathbf{x} = \mathbf{x}^*(\lambda\mathbf{x}) = \lambda\mathbf{x}^*\mathbf{x}$ and this can only be real if λ is real.

(ii) Suppose that $\mathbf{x}$ and $\mathbf{y}$ are eigenvectors with respective eigenvalues λ and μ. Then

$$\lambda\mathbf{y}^T\mathbf{x} = \mathbf{y}^T(\lambda\mathbf{x}) = \mathbf{y}^TA\mathbf{x} = \mathbf{y}^TA^T\mathbf{x} = (A\mathbf{y})^T\mathbf{x} = (\mu\mathbf{y})^T\mathbf{x} = \mu\mathbf{y}^T\mathbf{x}.$$

If $\lambda \neq \mu$ this means that $\mathbf{y}^T\mathbf{x} = 0$, hence they are at right angles to each other.

The spectral radius

Definition 5.4 *The* **spectrum** *of a matrix is the set of its eigenvalues and is written as $\sigma(A)$. That is,*

$$\sigma(A) = \{\lambda : \det(A - \lambda I) = 0\}.$$

The **spectral radius** *of A is defined to be the modulus of its largest eigenvalue and is written $\rho(A)$. That is,*

$$\rho(A) = \max_{\lambda \in \sigma(A)} |\lambda|.$$

The spectral radius of a matrix can be a particularly useful measure of the properties of a matrix and we will discuss how we can estimate it later in this chapter.

Suppose $A\mathbf{x} = \lambda\mathbf{x}$. Then $A^k\mathbf{x} = A^{k-1}(A\mathbf{x}) = \lambda A^{k-1}\mathbf{x}$. This identity forms the basis of an inductive proof that if λ is an eigenvalue of A then λ^k is an eigenvalue of A^k. Hence $\rho(A^k) = \rho(A)^k$. If $\rho(A) < 1$ then $\rho(A^k) \to 0$ as $k \to \infty$ and we can conclude that $A^k \to O$, too (the details are left to the reader). If $\rho(A) > 1$ we can show that A^k diverges. If $\rho(A) = 1$ then A^k may or may not converge. We will study matrix powers in more detail in chapter 12.

We already know that for real matrices, $\rho(A^T) = \rho(A)$. For complex matrices we know that λ is an eigenvalue of A if and only if $\bar{\lambda}$ is an eigenvalue of A^* and so, again, $\rho(A^*) = \rho(A)$. Furthermore, if A is nonsingular and $A\mathbf{x} = \lambda\mathbf{x}$ then $A^{-1}\mathbf{x} = \lambda^{-1}\mathbf{x}$, and we can conclude that

$$\rho(A^{-1}) = \max_{\lambda \in \sigma(A)} \frac{1}{|\lambda|}.$$

Similarity transforms

When it comes to utilizing spectral information to analyse networks, we will want to look at both eigenvalues and eigenvectors. Furthermore, the behaviour of matrix functions is intimately tied to the interplay of eigenvalues and eigenvectors. It is worthwhile, then, looking at the basic tools for understanding this interplay. Similarity transformations are of utmost importance in this regard.

Definition 5.5 *If X is a nonsingular matrix then the mapping*

$$S_X : A \rightarrow X^{-1}AX$$

is called a **similarity transformation.**

S_X *is an* **orthogonal similarity transformation** *if X is orthogonal and a* **unitary similarity transformation** *if X is unitary.*

Let $Ax = \lambda x$ and $B = C^{-1}AC$, then

$$BC^{-1}x = (C^{-1}AC)C^{-1}x = C^{-1}Ax = \lambda C^{-1}x.$$

So x is an eigenvector of A with eigenvalue λ iff $C^{-1}x$ is an eigenvector of B with eigenvalue λ. A 1-1 correspondence exists between the eigenpairs of A and B.

Example 5.2

Let

$$A = \begin{bmatrix} -5 & 1 & 3 \\ 18 & 2 & -9 \\ -20 & 2 & 11 \end{bmatrix}, \quad D = \begin{bmatrix} 1 & 0 & 0 \\ 0 & 2 & 0 \\ 0 & 0 & 5 \end{bmatrix}, \quad X = \begin{bmatrix} 1 & 1 & 0 \\ 0 & 1 & 3 \\ 2 & 2 & -1 \end{bmatrix}.$$

We can show that $A = XDX^{-1}$, given that

$$X^{-1} = \begin{bmatrix} 7 & -1 & -3 \\ -6 & 1 & 3 \\ 2 & 0 & -1 \end{bmatrix}.$$

Clearly D has eigenvalues 1, 2, and 5 with eigenvectors corresponding to the columns of I. It is straightforward to show that A has the same eigenvalues as D and the eigenvectors of A are the columns of X.

This example shows that under an appropriate similarity transformation, the eigenstructure of a matrix becomes obvious. The question remains as to how simple a form a matrix can be reduced to with similarity transformations.

5.2.1 The spectral theorem

In general, it is possible to find a similarity transformation for any $n \times n$ matrix A such that

$$X^{-1}AX = \mathcal{J},$$

where $\mathcal{J}$ is a block diagonal matrix $\mathcal{J} = \mathrm{diag}(\mathcal{J}_1, \ldots, \mathcal{J}_l)$ whose ith block is

$$
\mathcal{J}_i =
\begin{bmatrix}
\lambda_i & 1 & & \\
& \lambda_i & \ddots & \\
& & \ddots & 1 \\
& & & \lambda_i
\end{bmatrix},
$$

and λ_i is an eigenvalue of A. The matrix $\mathcal{J}$ is known as the *Jordan canonical form* of A and the representation $A = X\mathcal{J}X^{-1}$ is known as the Jordan decomposition.

If all the diagonal blocks are 1×1 then the matrix is called *simple*. Matrices that are not simple are *defective* and do not have a set of n linearly independent eigenvectors meaning that the algebraic multiplicity does not match the geometric multiplicity for at least one eigenvalue. Almost every matrix we deal with will be simple; certainly this is the case for simple networks, but defective matrices can arise for directed networks.

Example 5.3

The defective matrix

$$
A =
\begin{bmatrix}
-1 & 8 & 0 \\
0 & 3 & 1 \\
1 & -4 & 1
\end{bmatrix}
$$

has a single eigenvalue $\lambda = 1$. Its Jordan canonical form is

$$
\mathcal{J} =
\begin{bmatrix}
1 & 1 & 0 \\
0 & 1 & 1 \\
0 & 0 & 1
\end{bmatrix}.
$$

If

$$
X =
\begin{bmatrix}
4 & -2 & -7 \\
1 & 0 & -2 \\
-2 & 1 & 4
\end{bmatrix}
$$

then $A = XJX^{-1}$. The first column of X is the eigenvector of A and it is the only eigenvector of the matrix. The second and third columns satisfy the equations

$$(A-I)\mathbf{x}_{i+1} = \mathbf{x}_i, \qquad i = 1,2.$$

The diagonal blocks in the Jordan canonical form of a matrix are unique, hence two matrices are similar if and only if their Jordan canonical forms have the same diagonal blocks.

The Jordan decomposition is of great theoretical use but in practice it can be difficult to compute accurately as it can be highly sensitive to small perturbations. An alternative is to use the Schur decomposition.

Problem 5.2

Show that for any $n \times n$ matrix A there is a unitary matrix U and an upper-triangular matrix T such that

$$A = UTU^*. \tag{5.1}$$

This can be established by induction. The result is trivial for $n = 1$ ($U = 1$, $T = A$). Let us assume it holds for $n = k$.

Let A be a $(k+1) \times (k+1)$ matrix and choose one of its eigenvalues, λ with associated eigenvector $\mathbf{x}$ (where $\|\mathbf{x}\|_2 = 1$). Let X be a unitary matrix whose first column is $\mathbf{x}$ (can you describe how to find one?) and write $X = \begin{bmatrix} \mathbf{x} & W \end{bmatrix}$. Note that $W^*\mathbf{x} = 0$. Now,

$$X^*AX = \begin{bmatrix} \mathbf{x}^* \\ W^* \end{bmatrix} A \begin{bmatrix} \mathbf{x} & W \end{bmatrix} = \begin{bmatrix} \mathbf{x}^*A\mathbf{x} & \mathbf{x}^*AW \\ W^*A\mathbf{x} & W^*AW \end{bmatrix},$$

and since $A\mathbf{x} = \lambda\mathbf{x}$ it follows that $\mathbf{x}^*A\mathbf{x} = \lambda$ and $W^*A\mathbf{x} = \lambda W^*\mathbf{x} = 0$, so

$$X^*AX = \begin{bmatrix} \lambda & \mathbf{x}^*AW \\ 0 & W^*AW \end{bmatrix}.$$

Now W^*AW is a $k \times k$ matrix and so by the inductive hypothesis has a Schur decomposition VT_1V^*. Define

$$\widehat{V} = \begin{bmatrix} 1 & 0 \\ 0 & V \end{bmatrix}.$$

Then $\widehat{V}$ is unitary (why?) and

$$\widehat{V}^*X^*AX\widehat{V} = \begin{bmatrix} 1 & 0 \\ 0 & V^* \end{bmatrix}\begin{bmatrix} \lambda & \mathbf{x}^*AW \\ 0 & VT_1V^* \end{bmatrix}\begin{bmatrix} 1 & 0 \\ 0 & V \end{bmatrix} = \begin{bmatrix} \lambda & \mathbf{b}^* \\ 0 & T_1 \end{bmatrix},$$

(for some vector $\mathbf{b}$) which is upper triangular. Call this matrix $\widehat{T}$ and let $\widehat{U} = X\widehat{V}$, which is unitary (why?). $A = \widehat{U}\widehat{T}\widehat{U}^*$ is a Schur decomposition of A.

The matrix T in (5.1) is known as the *Schur canonical form* and UTU^* is the *Schur decomposition* of A.

If T is triangular,

$$\det(T) = \prod_{i=1}^{n} t_{ii},$$

and its characteristic polynomial is

$$\det(T - \lambda I) = (t_{11} - \lambda)(t_{22} - \lambda) \cdots (t_{nn} - \lambda).$$

Hence the eigenvalues of a triangular matrix are its diagonal elements. We do not need to transform a matrix to diagonal form to find its eigenvalues—just triangular form, computation of the eigenvectors is a further step, though. However if T is not only triangular but diagonal, the eigenvectors are trivial to find.

For a general real matrix, the Schur decomposition can be complex. But for most of our analysis we will be working with symmetric matrices for which the Schur decomposition simplifies and we arrive at the so-called *spectral theorem*.

Theorem 5.1 *If $A \in \mathbb{R}^{n \times n}$ is symmetric then there is an orthogonal matrix Q and a real diagonal matrix D such that*

$$A = QDQ^T \tag{5.2}$$

where $D = \mathrm{diag}(\lambda_1, \lambda_2, \ldots, \lambda_n)$ contains the eigenvalues of A and the columns of Q are the corresponding eigenvectors.

The spectral theorem is a direct consequence of the existence of a Schur decomposition. Suppose A has the Schur decomposition UTU^*. Then, since A is symmetric,

$$T^* = (U^*AU)^* = U^*A^*U = U^*AU = T,$$

hence T is a diagonal matrix, its entries must be the eigenvalues of A, UTU^* is the Jordan decomposition of A and the columns of U are the eigenvectors of A. But we know the eigenvalues and eigenvectors of a real symmetric matrix are real, so $D = T$ and $Q = U$.

So if A is symmetric, its eigenvectors are just as easy to determine as the eigenvalues once we have found the Schur decomposition. In this case the Schur decomposition is identical to the Jordan decomposition.

5.2.2 Gershgorin's theorem

Before we start to look for the eigenvalues of a matrix, it is useful to have some idea of where they lie in the complex plane. The following theorem gives us some useful information.

Theorem 5.2 (Gershgorin) *Let $A \in \mathbb{C}^{n \times n}$ and suppose that $X^{-1}AX = D + F$ where $D = \mathrm{diag}(d_1, \ldots, d_n)$ and F has no nonzero diagonal entries. Then the eigenvalues of A lie in the union of the discs $\Delta_1, \Delta_2, \ldots, \Delta_n$ where*

$$\Delta_i = \left\{ z \in \mathbb{C} : |z - d_i| \leq \sum_{j=1}^{n} |f_{ij}| \right\}.$$

If we pick X carefully we can often get tight bounds on the locations of the eigenvalues. Simple choices of X can also be useful. Note that if the discs are disjoint they each contain a single eigenvalue of A.

Example 5.4

Suppose $X = I$, then

$$\Delta_i = \left\{ z \in \mathbb{C} : |z - a_{ii}| \leq \sum_{j \neq i} |a_{ij}| \right\}.$$

For example, if

$$A = \begin{bmatrix} 10 & 2 & 3 \\ -1 & 0 & 2 \\ 1 & -2 & 1 \end{bmatrix}$$

then

$$\Delta_1 = \{z : |z - 10| \leq 5\}$$
$$\Delta_2 = \{z : |z| \leq 3\}$$
$$\Delta_3 = \{z : |z - 1| \leq 3\}.$$

The actual eigenvalues of A are 10.226 and $0.3870 \pm 2.216i$.

If we choose $X = \begin{bmatrix} 6 & 0 & 0 \\ 0 & 1 & 0 \\ 0 & 0 & 1 \end{bmatrix}$ then the discs are

$$\Delta_1 = \{z : |z - 10| \leq 5/6\}$$
$$\Delta_2 = \{z : |z| \leq 8\}$$
$$\Delta_3 = \{z : |z - 1| \leq 8\}$$

and we can give a much better estimate of the largest eigenvalue, as illustrated in Figure 5.1.

continued

Example 5.4 *continued*

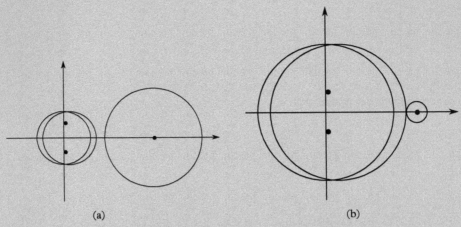

Figure 5.1 *Gershgorin discs for (a) X = I (b) X = diag(6, 1, 1). The eigenvalues are indicated by black circles*

If we want to estimate eigenvalues we can also make use of the *Rayleigh quotient*.

Definition 5.6   *Given $A \in \mathbb{R}^{n \times n}$ and $\mathbf{x} \in \mathbb{R}^n$ the* **Rayleigh quotient** *associated with A and $\mathbf{x}$ is $\dfrac{\mathbf{x}^T A \mathbf{x}}{\mathbf{x}^T \mathbf{x}}$.*

The Rayleigh quotient appears in a number of guises in applied mathematics. If $\mathbf{x}$ is an eigenvector (with associated eigenvalue λ) then

$$\frac{\mathbf{x}^T A \mathbf{x}}{\mathbf{x}^T \mathbf{x}} = \frac{\mathbf{x}^T \lambda \mathbf{x}}{\mathbf{x}^T \mathbf{x}} = \lambda.$$

The range of Rayleigh quotients for a given matrix will thus include the whole spectrum (if that spectrum is real). Bounds on this range can be used to bound the spectral radius (which is an upper bound on the size of Rayleigh quotients of a given matrix).

5.2.3   Perron–Frobenius theorem

Adjacency matrices are *nonnegative*. That is, all of their entries are greater than or equal to zero. We use the inequality $A \geq 0$ to denote that A is nonnegative. We can say something concrete about the largest eigenvalue of nonnegative matrices which will be useful when we analyse networks. In terms of adjacency matrices, the key results are dependent on the connectivity of the underlying network. They follow from results which were first proved for *positive matrices*. If A is a positive matrix then we write $A > 0$.

Theorem 5.3 (Perron) *Suppose $A \in \mathbb{R}^{n \times n}$ and $A > 0$. Then A has an eigenvalue λ that satisfies the following properties.*

1. *$\lambda = \rho(A)$.*
2. *If $\mu \in \sigma(A)$ and $\mu \neq \lambda$ then $|\mu| < \lambda$.*
3. *λ has algebraic multiplicity 1.*
4. *If $A\mathbf{y} = \lambda\mathbf{y}$ then $\mathbf{y} = \alpha\mathbf{x}$ where $\mathbf{x} > 0$ and $\alpha \in \mathbb{C}$.*

The proof of Perron's theorem can be found in many standard linear algebra texts. The details are rather intricate and we omit them here, but it would be remiss of us not to give a flavour of what the proof involves.

Problem 5.3
Show that if $A > 0$ then the following are true.

1. $A^k > 0$ for every finite k.
2. $\rho(A) > 0$.
3. If $\mathbf{x} \geq 0$ then $A\mathbf{x} \geq 0$ with equality if and only if $\mathbf{x} = 0$.

1. By induction. $A > 0$. Suppose $B = A^k > 0$. Then the (i,j)th entry of A^{k+1} is $\sum_{k=1}^{n} b_{ik}a_{kj}$ and the result follows since every term in this sum is positive.
2. If $\rho(A) = 0$ then $A^k = O$ for $k \geq n$. But we have just shown that $A^k > 0$ for a positive matrix.
3. The kth entry of $A\mathbf{x}$ is $\sum_{k=1}^{n} a_{ik}x_k$. Every term is nonnegative and the sum can only equal zero if every element is zero.

For nonnegative matrices, point one of the Perron theorem is also true. But the uniqueness of the largest eigenvalue is not guaranteed.

Examples 5.5

(i) $A = \begin{bmatrix} 0 & 1 \\ 1 & 0 \end{bmatrix}$ is nonnegative and has eigenvalues ± 1, showing point two of the Perron theorem does not hold for all nonnegative matrices.

(ii) Suppose $A > 0$ and that $A\mathbf{x} = \rho(A)\mathbf{x}$ where $\mathbf{x} > 0$. Now let $B = \begin{bmatrix} A & O \\ O & A \end{bmatrix}$. Then it should be obvious that $\rho(B) = \rho(A)$ and

$$B\begin{bmatrix} \mathbf{x} \\ 0 \end{bmatrix} = \rho(B)\begin{bmatrix} \mathbf{x} \\ 0 \end{bmatrix}, \quad B\begin{bmatrix} \mathbf{x} \\ -\mathbf{x} \end{bmatrix} = \rho(A)\begin{bmatrix} \mathbf{x} \\ -\mathbf{x} \end{bmatrix},$$

proving that points three and four of the Perron theorem do not hold for all nonnegative matrices.

(iii) We could have used $B = I$ in the last example.

Many nonnegative matrices do share all the key properties of positive matrices. Whether they do or not depends on the pattern of zeros in the matrix which for adjacency matrices, as previously stated, can be related to *connectivity* in networks.

Definition 5.7 $A \in \mathbb{R}^{n \times n}$ is **reducible** *if there exists a permutation P such that*

$$A = P \begin{bmatrix} X & Y \\ O & Z \end{bmatrix} P^T$$

where X and Z are both square. $A \in \mathbb{R}^{n \times n}$ is **fully decomposable** *if there are permutations P and Q such that*

$$A = P \begin{bmatrix} X & Y \\ O & Z \end{bmatrix} Q.$$

If a square matrix is not reducible then it is **irreducible**. *If it is not fully decomposable then it is* **fully indecomposable**.

Theorem 5.4 (Perron–Frobenius)   *If A is fully indecomposable and nonnegative then the properties listed in the Perron theorem still hold. If it is irreducible then properties 1, 3, and 4 are guaranteed to hold.*

Perron's theorem and the Perron–Frobenius theorem can be generalized in many ways to cover much more exotic linear operators.

Notice that if A is symmetric and reducible then it can be permuted into the block diagonal form $\begin{bmatrix} X & O \\ O & Z \end{bmatrix}$. Essentially, X and Z are completely independent of each other. If either of these matrices is reducible it can also be permuted into block diagonal form, and so on, and hence any symmetric matrix can be written in the form

$$A = P\mathrm{diag}(A_1, A_2, \ldots, A_k)P^T$$

where each of the diagonal blocks is irreducible. The Perron–Frobenius theorem can then be applied to each block in turn.

Notice that using reducibility we can restate the condition established in Chapter 2 that characterizes connectivity. A network, G, is connected if and only if its adjacency matrix, A, is irreducible.

If a network has more than one connected component then its adjacency matrix must be reducible. In particular, using the argument given above, if the network is undirected we can permute its adjacency matrix into a block diagonal form in which each block is irreducible.

FURTHER READING

Horn, R.A. and Johnson, C.R., *Matrix Analysis*, Cambridge University Press, 2012.
Meyer, C.D., *Matrix Analysis and Applied Linear Algebra*, SIAM, 2000.
Strang, G., *Linear Algebra and Its Applications*, Brooks/Cole, 2004.

Spectra of Adjacency Matrices

In this chapter

Diverse physical phenomena can be understood by studying their spectral properties. An understanding of the harmonies of music can be developed by looking at characteristic frequencies that can be viewed as eigenfunctions; and astronomers can predict the chemical composition of unimaginably distant galaxies from the spectra of the electromagnetic radiation they emit.

In this chapter we look at some ways to define the spectrum of a network and what we can infer from the resulting eigenvalues. We will only consider undirected networks, which allows us to take advantage of some powerful tools from matrix algebra.

6.1 Motivation

The obvious place to start when looking for the spectrum of a network is the adjacency matrix. For now, we will focus on simple networks. Since the adjacency matrix is symmetric, the eigenvalues are real (by the spectral theorem) and since it is nonnegative, its largest eigenvalue is real and positive (by the Perron–Frobenius theorem). We can compare networks through the spectra of their adjacency matrices but we can also calculate some useful network statistics from them, too. We will assume that the spectrum of the adjacency matrix A is ordered so that

$$\lambda_1 \geq \lambda_2 \geq \cdots \geq \lambda_n.$$

Since A is symmetric and the eigenvalues are real, such an ordering is possible.

6.2 Spectral analysis of simple networks

The spectrum of the adjacency matrix is an example of a *graph invariant*. However we decide to label a network, the eigenvalues we compute will be the same. This is simple to see. Relabelling is equivalent to applying a similarity transformation induced by a permutation matrix, P, and

$$\det(PAP^T - \lambda I) = \det(P(A - \lambda I)P^T) = \det(A - \lambda I)\det(P)\det(P^T)$$

and hence the zeros of the c.p. are unchanged by the permutation.

A network is not uniquely defined by its spectrum—there are a number of famous results in graph theory on isospectral graphs (i.e. graphs with exactly the same eigenvalues) which are otherwise unrelated—but certain properties can still be inferred. The spectra of some highly structured networks can be derived analytically through a variety of straightforward linear algebra techniques.

Examples 6.1

(i) The adjacency matrix of the complete network[1] K_n is $A = E - I$. E has a zero eigenvalue of algebraic multiplicity $n-1$ (why?), so A has an eigenvalue of -1 with algebraic multiplicity $n-1$. Since $A\mathbf{e} = (n-1)\mathbf{e}$, the remaining eigenvalue is $n-1$. Thus

$$\det(\lambda I - A) = (\lambda + 1)^{n-1}(\lambda - n + 1).$$

(ii) The cycle graph, C_n, has adjacency matrix

$$A = \begin{bmatrix} 0 & 1 & 0 & \cdots & 0 & 1 \\ 1 & 0 & 1 & 0 & \cdots & 0 \\ 0 & 1 & 0 & \ddots & & 0 \\ \vdots & 0 & 1 & \ddots & \ddots & \vdots \\ 0 & \vdots & & \ddots & 0 & 1 \\ 1 & 0 & \cdots & 0 & 1 & 0 \end{bmatrix}.$$

Let $\omega^n = 1$ and $\mathbf{v} = \begin{bmatrix} 1 & \omega & \omega^2 & \cdots & \omega^{n-1} \end{bmatrix}^T$. Then, since $\omega^{n-1} = \omega^{-1}$,

$$A\mathbf{v} = \begin{bmatrix} \omega + \omega^{-1} \\ 1 + \omega^2 \\ \omega + \omega^3 \\ \vdots \\ \omega^{n-3} + \omega^{n-1} \\ 1 + \omega^{n-2} \end{bmatrix} = (\omega + \omega^{-1})\mathbf{v}.$$

Thus $\omega + \omega^{-1} = 2\mathbf{Re}\,\omega$ is an eigenvalue of A for any root of unity, ω. This means the spectrum of C_n is

$$\left\{ 2\cos\left(\frac{2\pi j}{n}\right), j = 1, \ldots, n \right\}.$$

All the eigenvalues corresponding to complex ω have algebraic multiplicity 2 since ω^j and ω^{n-j} share the same real part.

continued

[1] Throughout this book $\mathbf{e}$ denotes a vector of ones and E a matrix of ones. Their dimensions will vary but should be readily understood from the context in which they appear.

Examples 6.1 *continued*

(iii) It can be shown that each repeated eigenvalues of C_{2n} is an eigenvalue of the path graph. This can be explained by looking closely at the eigenvectors of C_{2n}, which we will do in Section 6.4. Accordingly, the spectrum of P_{n-2} is

$$\left\{ 2\cos\left(\frac{\pi j}{n}\right), j = 1, \ldots, n-1 \right\}.$$

(iv) If G is bipartite then its adjacency matrix can be permuted to the form

$$A = \begin{bmatrix} O & B \\ B^T & O \end{bmatrix}.$$

Suppose $\mathbf{v}$ is an eigenvector of A. Partition it as we have A so $\mathbf{v} = \begin{bmatrix} \mathbf{x} \\ \mathbf{y} \end{bmatrix}$. Now $A\mathbf{v} = \lambda\mathbf{v}$ and

$$A\mathbf{v} = \begin{bmatrix} O & B \\ B^T & O \end{bmatrix}\begin{bmatrix} \mathbf{x} \\ \mathbf{y} \end{bmatrix} = \begin{bmatrix} B\mathbf{y} \\ B^T\mathbf{x} \end{bmatrix},$$

so $B\mathbf{y} = \lambda\mathbf{x}$ and $B^T\mathbf{x} = \lambda\mathbf{y}$. Hence

$$A\begin{bmatrix} \mathbf{x} \\ -\mathbf{y} \end{bmatrix} = \begin{bmatrix} -B\mathbf{y} \\ B^T\mathbf{x} \end{bmatrix} = \begin{bmatrix} -\lambda\mathbf{x} \\ \lambda\mathbf{y} \end{bmatrix} = -\lambda\begin{bmatrix} \mathbf{x} \\ -\mathbf{y} \end{bmatrix}.$$

We conclude that if $\lambda \in \sigma(A)$ then so is $-\lambda$ and the spectrum of a bipartite network is centred around zero. In particular, a bipartite network with an odd number of nodes must have a zero eigenvalue.

(v) Let E_{mn} represent an $m \times n$ array of ones. Then the adjacency matrix of the complete bipartite network $K_{m,n}$ can be written

$$A = \begin{bmatrix} O & E_{mn} \\ E_{mn}^T & O \end{bmatrix}.$$

Suppose that $\begin{bmatrix} \mathbf{x} \\ \mathbf{y} \end{bmatrix}$ is an eigenvector of A (where $\mathbf{x}$ has m components) with eigenvalue λ. Then

$$\lambda\begin{bmatrix} \mathbf{x} \\ \mathbf{y} \end{bmatrix} = A\begin{bmatrix} \mathbf{x} \\ \mathbf{y} \end{bmatrix} = \begin{bmatrix} E_{mn}\mathbf{y} \\ E_{mn}^T\mathbf{x} \end{bmatrix},$$

so

$$E_{mn}E_{mn}^T\mathbf{x} = \lambda E_{mn}\mathbf{y} = \lambda^2\mathbf{x}.$$

But $E_{mn}E_{mn}^T = nE$, where E is an $m \times m$ matrix of ones. From our analysis of the complete graph, we know that the spectrum of nE is $\{mn, 0\}$ and so the eigenvalues of K_{mn} are $\pm\sqrt{mn}$ and 0. We get a similar result if we consider $E_{mn}^T E_{mn}\mathbf{y}$ and if we account for all the copies of the zero eigenvalue we find that it has algebraic multiplicity of $m + n - 2$. Hence,

$$\det(\lambda I - A) = \lambda^{m+n-2}(\lambda^2 - mn).$$

(vi) If G is disconnected then its spectrum is simply the union of the spectra of the individual components.

(vii) If the adjacency matrix of G has characteristic polynomial

$$\det(\lambda I - A) = (\lambda - \lambda_1)^{p_1}(\lambda - \lambda_1)^{p_2}\cdots(\lambda - \lambda_k)^{p_k}$$

then it is common practice in network theory to write its spectrum as

$$\sigma(G) = \{[\lambda_1]^{p_1}, [\lambda_2]^{p_2}, \ldots, [\lambda_k]^{p_k}\}.$$

For general networks we can either compute (parts of) the spectrum numerically or estimate certain eigenvalues. The most useful eigenvalues to have information about are λ_1, λ_2, and λ_n (the most negative eigenvalue). Since A is symmetric and nonnegative we can deduce several things about λ_1 and λ_n.

By Gershgorin's theorem, the spectrum lies in discs centred on the diagonal elements of A and have radius equal to the sum of off diagonal elements in a row. Note that since the eigenvalues are real, this means all eigenvalues lie in the interval $[-n + 1, n - 1]$. By the Perron–Frobenius theorem, the largest eigenvalue is nonnegative so $0 \le \lambda_1 \le n - 1$ and by symmetry, $\lambda_1 = 0$ if and only if $A = O$. Note that the adjacency matrices of all networks apart from that of N_n have negative eigenvalues.

Another consequence of the Perron–Frobenius theorem is that unless A is reducible, $|\lambda_n| < \lambda_1$. We saw in an earlier example that the spectrum of a bipartite network is symmetric about zero, and hence $|\lambda_n| = \lambda_1$. The only other case in which A is reducible for a simple network, G, is if G has several components. And unless one of these components is bipartite, the smallest eigenvalue is guaranteed to satisfy $|\lambda_n| < \lambda_1$, too.

We can get a lower bound on the biggest eigenvalue using the ratio of edges to nodes in a network. Since $\lambda_1 = \rho(A)$ is an upper bound on the size of Rayleigh quotients, for any $\mathbf{x} \ne 0$, $\lambda_1 \ge \dfrac{\mathbf{x}^T A \mathbf{x}}{\mathbf{x}^T \mathbf{x}}$. Letting $\mathbf{x} = \mathbf{e}$ gives

$$\lambda_1 \geq \frac{e^T A e}{e^T e} = \frac{2m}{n}.$$

We can get tighter bounds on eigenvalues by using additional information about the network.

Problem 6.1

Show that if $k_{\max}$ is the maximal degree of any node of a simple network with adjacency matrix A then $\lambda_1 \geq \sqrt{k_{\max}}$.

Suppose node k has the highest degree and define $\mathbf{x}$ so

$$x_i = \begin{cases} \sqrt{k_{\max}}, & i = k, \\ a_{ik}, & i \neq k, \end{cases}$$

and let $\mathbf{y} = A\mathbf{x}$.

We can get lower bounds on the components of $\mathbf{y}$. They are all nonnegative and if $a_{ik} = 1$, $y_i \geq a_{ik} x_k = \sqrt{k_{\max}}$. Noting that A has a zero diagonal,

$$y_k = \sum_j a_{kj} x_j = \sum_j a_{kj}^2 = k_{\max}.$$

Since $\mathbf{x}^T \mathbf{x} = 2k_{\max}$,

$$\lambda_1 \geq \frac{\mathbf{x}^T A \mathbf{x}}{\mathbf{x}^T \mathbf{x}} \geq \frac{k_{\max}\sqrt{k_{\max}} + \sqrt{k_{\max}}k_{\max}}{2k_{\max}} = \sqrt{k_{\max}}.$$

Note that the second largest eigenvalue, λ_2, is harder to estimate accurately. The *spectral gap* $\lambda_1 - \lambda_2$ can give useful insight into a network but we will not discuss this further.

6.3 Spectra and structure

In Section 6.2 we showed how we could use features of the network to estimate eigenvalues. We can work the other way round, too: certain spectral distributions guarantee the existence of particular network structures, and we can use the eigenvalues to enumerate these.

Suppose that $G(V, E)$ is a simple network whose adjacency matrix A has characteristic polynomial

$$\det(\lambda I - A) = c_0 + c_1\lambda + \cdots + c_{n-1}\lambda^{n-1} + c_n\lambda^n.$$

As previously stated, the coefficients of the c.p. are related to the sum of the principal minors of A. Since A is simple its diagonal is zero and hence so is that of any principal submatrix. All nonzeros in an adjacency matrix are ones and all principal submatrices must be symmetric. A 1×1 principal minor of A is simply a diagonal element hence $c_{n-1} = 0$.

The possible 2×2 principal submatrices are $\begin{bmatrix} 0 & 0 \\ 0 & 0 \end{bmatrix}$ and $\begin{bmatrix} 0 & 1 \\ 1 & 0 \end{bmatrix}$. Only the second of these has a nonzero determinant (namely -1) and there is one of these principal submatrices for every edge in A. Thus $-c_{n-2}$ equals the number of edges in G.

The non-trivial 3×3 principal submatrices are

$$\begin{bmatrix} 0 & 1 & 0 \\ 1 & 0 & 0 \\ 0 & 0 & 0 \end{bmatrix}, \quad \begin{bmatrix} 0 & 1 & 1 \\ 1 & 0 & 0 \\ 1 & 0 & 0 \end{bmatrix}, \quad \text{and} \quad \begin{bmatrix} 0 & 1 & 1 \\ 1 & 0 & 1 \\ 1 & 1 & 0 \end{bmatrix}.$$

Only the third of these has a nonzero determinant (it is two). Such a submatrix corresponds to a triangle in G and so $-c_{n-3}$ counts twice the number of triangles in G.

We can use the spectrum of a network to count walks. In particular, since the entries of A^p denote the number of walks of length p between nodes, the number of closed walks lies along the diagonal and hence the sum of the closed walks of length p is

$$\mathrm{tr}(A^p) = \sum_{i=1}^{n} \lambda_i^p, \tag{6.1}$$

where the λ_i are the eigenvalues of A.

We can use this result as an alternative method for counting edges and triangles. Every closed walk of length two represents a trip along an edge and back. Each edge is involved in two such trips. Every closed walk of length three represents a trip around a triangle. Each triangle is involved in six such walks. So just divide (6.1) by 2 or by 6 when $p = 2$ or 3 to calculate the right values. Things get a little trickier if we want to count the number of circuits of greater length, as we have to eliminate closed walks which are not paths.

Examples 6.2

(i) Recall that the c.p. of the complete graph is

$$\det(\lambda I - A) = (\lambda + 1)^{n-1}(\lambda - n + 1).$$

Expanding this polynomial gives

$$\det(\lambda I - A) = \lambda^n - \frac{n(n-1)}{2}\lambda^{n-2} - \frac{n(n-1)(n-2)}{3}\lambda^{n-3} + \cdots + (1-n).$$

continued

Examples 6.2 *continued*

One can readily confirm that the coefficients of λ^{n-2} and λ^{n-3} match the expected values in terms of edges and triangles.

(ii) The number of edges in the cycle graph C_n is n and there are no triangles for $n \geq 4$, thus its characteristic polynomial is

$$\lambda^n - n\lambda^{n-2} + c_n\lambda^{n-3} + \cdots$$

where $c_3 = -2$ and $c_n = 0$ if $n > 3$.

(iii) The path graph, P_n, has n edges and no triangles, so the c.p. of its adjacency matrix is

$$\det(\lambda I - A) = \lambda^{n+1} - n\lambda^{n-1} + 0\lambda^{n-2} + \cdots.$$

(iv) Since a bipartite network has no triangles, the c_{n-3} coefficient in the c.p. is zero.

(v) Suppose that the spectrum of the adjacency matrix of a network is symmetric about zero. Then if p is odd,

$$\text{tr}(A^p) = \sum_{i=1}^{n} \lambda_i^p = 0,$$

so there are no closed walks in the network of odd length, meaning that it is bipartite. This is the converse of the result we established in Example 6.1(iv).

We can also use (6.1) to predict whether a network contains certain features.

Problem 6.2

Suppose $G = (V, E)$ has m edges. Show that if $\lambda_1 > \sqrt{m}$, then G contains a triangle.

Since $m \geq 0$, $\lambda_1 > 0$. From (6.1), $2\lambda_1^3 > 2\lambda_1 m \geq \lambda_1 \sum_{i=1}^{n} \lambda_i^2 \geq \sum_{i=1}^{n} |\lambda_i^3|$, so $\lambda_1^3 > \sum_{i=2}^{n} |\lambda_i^3|$.

If t is the number of triangles then $6t = \sum_{i=1}^{n} \lambda_i^3 \geq \lambda_1^3 - \sum_{i=2}^{n} |\lambda_i^3| > 0$.

Since t must take an integer value, we have established the existence of at least one triangle.

While this may not be the most practical of results, it illustrates the point that if we know or can estimate only limited parts of the spectrum, we may be able to determine many characteristics of the network.

Example 6.3

Suppose that G is a connected k-regular network with n nodes and that its adjacency matrix has only four distinct eigenvalues, namely,

$$\lambda_1 > \lambda_2 > \lambda_3 > \lambda_4,$$

such that

$$\sigma(G) = \left\{ [\lambda_1]^{p_1}, [\lambda_2]^{p_2}, [\lambda_3]^{p_3}, [\lambda_4]^{p_4} \right\}.$$

Since G is connected, $p_1 = 1$. And since $\sum p_i = n$ (algebraic multiplicities always sum to the dimension), we get

$$1 + p_2 + p_3 + p_4 = n.$$

The Gershgorin discs for the adjacency matrix, A, of a k-regular network are all of the form $|z| \leq k$ so $|\lambda_1| \leq k$ and since $A\mathbf{e} = k\mathbf{e}$ we know that $\lambda_1 = k$.

Since the sum of the eigenvalues of A is zero (since $\text{tr}(A) = 0$),

$$k + p_2\lambda_2 + p_3\lambda_3 + p_4\lambda_4 = 0.$$

Similarly,

$$k^2 + p_2\lambda_2^2 + p_3\lambda_3^2 + p_4\lambda_4^2 = \text{tr}(A^2) = nk,$$

since $\text{tr}(A^2)$ double counts the $nk/2$ edges of G.

Note also that the number of triangles, t, can be written

$$t = \frac{1}{6}\text{tr}(A^3) = \frac{1}{6}(k^3 + p_2\lambda_2^3 + p_3\lambda_3^3 + p_4\lambda_4^3).$$

Problem 6.3

Calculate the number of 4-cycles in the network described in Example 6.3 in terms of k, n, the eigenvalues and their multiplicities.

Given the constraints that we have imposed, it can be shown that the number of closed walks of a particular length is the same from any node in the network.

Suppose

$$u \to v \to w \to x \to u$$

is a closed walk of length four. Following the previous example, we can express the number of closed walks of length four as

$$\text{tr}(A^4) = k^4 + p_2\lambda_2^4 + p_3\lambda_3^4 + p_4\lambda_4^4.$$

In some of these, the nodes are not all distinct and so do not count towards the total of 4-cycles.

There are three types of walk of length four with duplicate nodes.

1. $u \to v \to u \to v \to u$
2. $u \to v \to u \to x \to u \quad (x \neq v)$
3. $u \to v \to x \to v \to u \quad (x \neq u)$

Given a node u, there are k choices for v and $k-1$ for x: any node adjacent to u (u) that is not v (u).

In total this gives k walks of type one and $k(k-1)$ of each of types two and three. We get the same number from any node in a k-regular graph, hence the total number of closed walks of length four which are not cycles is $nk(2k-1)$.

Each 4-cycle represents eight different closed walks: start from one of its nodes and move clockwise or anticlockwise.

Thus the total number of distinct 4-cycles is

$$\frac{1}{8}\left(k^4 + p_2\lambda_2^4 + p_3\lambda_3^4 + p_4\lambda_4^4 - nk(2k-1)\right).$$

Multiple eigenvalues appear to be much more common in adjacency matrices than in a matrix chosen at random. They often correspond to symmetries within the network (which is why it is possible for regular graphs to exist with very few distinct eigenvalues). A symmetry can be characterized as a relabelling of the nodes which leaves the adjacency matrix unchanged—a permutation P such that $P^T AP = A$. The possibilities for such a P are severely restricted if there are no repeated eigenvalues.

Example 6.4

Suppose that the adjacency matrix A has distinct eigenvalues and that $\mathbf{x}$ is an eigenvector of A such that $A\mathbf{x} = \lambda\mathbf{x}$.

If $P^T AP = A$ then $AP\mathbf{x} = PA\mathbf{x} = \lambda(P\mathbf{x})$ and so $P\mathbf{x}$ is also an eigenvector of A corresponding to the eigenvalue λ. This is only possible if $P\mathbf{x} = \pm\mathbf{x}$ which means that $P^2\mathbf{x} = \mathbf{x}$. Since this is true for all eigenvectors, $P^2 = I$ (and $P = P^T$).

The only permutations which have this property are the identity matrix and ones where pairs of nodes are swapped with each other.

6.4   Eigenvectors of the adjacency matrix

In classical graph theory there seem to be very few applications of eigenvectors, particularly those of the adjacency matrix, but when it comes to drawing out properties of complex networks, eigenvectors can prove to be a vital tool. For the most part, practitioners have used the eigenvectors of the *graph Laplacian*—which we introduce in Chapter 7—but the eigenvectors of the adjacency matrix have applications in areas such as community discovery (see Chapter 21). To do this, we will associate the ith component of an eigenvector with the ith node of the network and this value can be used as a quantifier.

The most frequently used eigenvector of the adjacency matrix is its principal one. Recall that in a nonnegative matrix, the eigenvector associated with λ_1 can be normalized so that all its entries are nonnegative. The entries can then be used to score the importance of a node; an idea we will make more concrete, in Chapter 15. In any case, we can use the eigenvector components to group together nodes with similar properties.

Examples 6.5

(i)   Consider a bipartite network with principal eigenvalue λ. From Example 6.1(iv), we know that the signs of the elements of the eigenvector associated with the eigenvalue $-\lambda$ can be used to divide the nodes of the network into its two parts.

(ii) We know that if ω is an nth root of 1 then

$$\mathbf{v} = \begin{bmatrix} 1 & \omega & \omega^2 & \dots & \omega^{n-1} \end{bmatrix}^T$$

is an eigenvector of C_n (see Example 6.1(ii)) with corresponding eigenvalue $2\cos(2\pi j/n)$ for some $j \in \{1, 2, \dots, n\}$. In this case the principal eigenvector is $\mathbf{e}$. All the nodes are identical in this network and so it is no surprise that they are indistinguishable. Note that $\mathbf{e}$ is an eigenvector of any regular network.

If n is even then C_n is bipartite and we can divide the network according to the signs of the elements of the eigenvector associated with the eigenvalue -2. In this case, the eigenvector is

$$\mathbf{v} = \begin{bmatrix} 1 & -1 & 1 & \dots & -1 \end{bmatrix}^T,$$

highlighting that neighbouring nodes in the cycle belong to different parts.

All the other eigenvalues of C_n are repeated, meaning that they are associated with two-dimensional eigenspaces. This makes it harder to use eigenvectors to differentiate elements. Orderings and splittings of nodes may vary according to the representative vectors we pick from the eigenspace.

(iii) We can exploit the eigenspaces of the repeated eigenvalues of the cycle graph C_{2n} to deduce the spectrum of the path graph P_{n-2}. First observe that removing diametrically positioned elements of C_{2n} gives us two copies of P_{n-2} (see Figure 6.1).

continued

Examples 6.5 *continued*

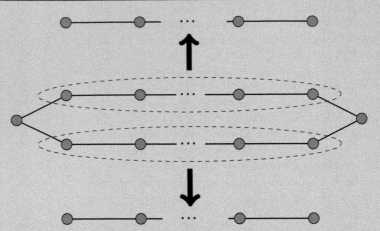

Figure 6.1 *Splitting the cycle graph C_{2n} into two copies of the path graph P_{n-2}*

Using Example 6.1(ii) we can write a basis for the two-dimensional eigenspaces in the form

$$\left\{ \begin{bmatrix} 1 \\ \omega \\ \omega^2 \\ \vdots \\ \omega^{2n-1} \end{bmatrix}, \begin{bmatrix} 1 \\ \omega^{-1} \\ \omega^{-2} \\ \vdots \\ \omega^{1-2n} \end{bmatrix} \right\},$$

where $\omega^{2n} = 1$.

Now choose such an ω, for which the eigenvalue is $\lambda = \omega + \omega^{-1}$, and call the basis elements $\mathbf{v}_1$ and $\mathbf{v}_2$, respectively. Note that the first and $(n+1)$th elements of $\mathbf{x} = \mathbf{v}_1 - \mathbf{v}_2$ are both zero.

Since $a_{1,n+1} = a_{n+1,1} = 0$ it follows that λ is an eigenvalue and $\mathbf{x}$ is an eigenvector of the matrix you obtain by replacing the first and $(n+1)$th rows and columns of A with zeros. For example, when $n = 4$ we end up with the matrix

$$\begin{bmatrix} 0 & 0 & 0 & 0 & 0 & 0 & 0 & 0 \\ 0 & 0 & 1 & 0 & 0 & 0 & 0 & 0 \\ 0 & 1 & 0 & 1 & 0 & 0 & 0 & 0 \\ 0 & 0 & 1 & 0 & 0 & 0 & 0 & 0 \\ 0 & 0 & 0 & 0 & 0 & 0 & 0 & 0 \\ 0 & 0 & 0 & 0 & 0 & 0 & 1 & 0 \\ 0 & 0 & 0 & 0 & 0 & 1 & 0 & 1 \\ 0 & 0 & 0 & 0 & 0 & 0 & 1 & 0 \end{bmatrix},$$

isolating two copies of the adjacency matrix of the path graph P_{n-2}. Thus each repeated eigenvalue of C_{2n} is an eigenvalue of P_{n-2}, as claimed in Example 6.1 (iii).

(iv) Consider the network in Figure 6.2.

The principal eigenvector of the adjacency matrix is

$$\begin{bmatrix} 0.106 & 0.044 & 0.128 & 0.219 & 0.177 & 0.186 & 0.140 \end{bmatrix}^T.$$

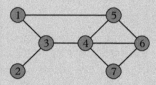

Figure 6.2 *A graph with a hub at node 4*

Notice that the largest element (the fourth) is associated with the node at the hub of the network and that the smallest element is associated with the most peripheral. The eigenvector of the most negative eigenvalue is

$$\begin{bmatrix} -0.172 & -0.091 & 0.206 & -0.202 & 0.183 & -0.039 & 0.107 \end{bmatrix}^T.$$

If we split the nodes according to the signs of the elements of this vector we find a grouping of nodes that points towards a bipartite split in the network we only need to remove the edge between nodes 4 and 6 to achieve this.

We will revisit applications of eigenvectors frequently in later chapters.

....

FURTHER READING

Biggs, N. *Algebraic Graph Theory*, Cambridge University Press, 1993.
Cvetković, D., Rowlinson, P., and Simić, S. *Eigenspaces of Graphs*, Cambridge University Press, 1997.
Cvetković, D., Rowlinson, P., and Simić, S. *An Introduction to the Theory of Graph Spectra*, Cambridge University Press, 2010.

The Network Laplacian

In this chapter

The adjacency matrix is not the only useful algebraic representation of a network, particularly when it comes to spectral analysis. We introduce the network Laplacian, which can be defined in terms of either the adjacency matrix or the incidence matrix. We look at the spectrum of the Laplacian, highlighting some pertinent properties and deriving some simple bounds on key eigenvalues. We give examples of the applications of the eigenvalues *and* eigenvectors of the Laplacian matrix. We will have more to say on these later.

7.1 The graph Laplacian

You may well have encountered the *Laplacian operator*, $\Delta = \nabla^2$ which plays a significant role in the theory and solution of partial differential equations. For example, solutions of the equation

$$\Delta u + \lambda u = 0$$

over a bounded region tell us about characteristic modes associated with vibrations in the region and Laplace's equation $\Delta u = 0$ can be related to equilibria in a system.

The *graph Laplacian* extends the idea of Δ to a discrete network. Starting from the definition of a derivative as the limit of differences,

$$f'(a) = \lim_{x \to a} \frac{f(x) - f(a)}{x - a},$$

we note that the *incidence matrix* lets us find differences between nodes in a network. Consider a function f which is assigned a value $f(x_i) = f_i$ at each node. Then the differences between incident nodes can be found by calculating $\mathbf{g} = B\mathbf{f}$, where B is the incidence matrix. To get the second difference at a node i, as we must to find Δ, we need to take the differences of all the first differences incident to i, namely $B^T\mathbf{g} = B^T B\mathbf{f}$. With a little more rigour (we need to choose appropriate denominators for our differences and make sure signs of differences are properly matched together) we can show that the matrix $L = B^T B$ is indeed a discrete analogue of the continuous Laplacian operator. We call it the *graph Laplacian* of the network G. The graph Laplacian can also be written in terms of the adjacency matrix. Note that for a network with n nodes and m edges, L is an $n \times n$ matrix and

$$l_{ij} = \sum_{k=1}^{m} b_{ki} b_{kj}.$$

Recall that row k of B contains only two nonzeros: a 1 and a -1 in the columns incident to the kth edge. There is a positive contribution to l_{ii} every time $b_{ki} = \pm 1$. That is, there is a positive contribution of 1 corresponding to every edge incident to i and hence l_{ii} is equal to the degree of node i. But if $i \neq j$ then $b_{ki} b_{kj} \neq 0$ if and only if i and j are adjacent in which case $b_{ki} b_{kj} = -1$. Therefore

$$L = D - A$$

where A is the adjacency matrix of G and $D = \mathrm{diag}(A\mathbf{e})$ is a diagonal matrix whose entries are the degrees of each node. This representation of L makes it easy to compute.

Examples 7.1

(i) Two networks are illustrated in Figure 7.1.

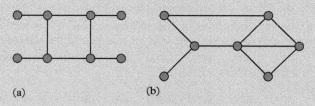

(a)   (b)

Figure 7.1 *Two simple networks*

Ordering nodes from left to right and top to bottom, their Laplacian matrices are

$$
\begin{bmatrix}
1 & -1 & 0 & 0 & 0 & 0 & 0 & 0 \\
0 & 1 & -1 & 0 & 0 & 0 & 0 & 0 \\
-1 & 0 & 3 & -1 & -1 & 0 & 0 & 0 \\
0 & -1 & -1 & 3 & 0 & -1 & 0 & 0 \\
0 & -1 & 0 & 0 & 3 & -1 & -1 & 0 \\
0 & 0 & 0 & -1 & -1 & 3 & 0 & -1 \\
0 & 0 & 0 & 0 & -1 & 0 & 1 & 0 \\
0 & 0 & 0 & 0 & 0 & -1 & 0 & 1
\end{bmatrix}
\text{ and }
\begin{bmatrix}
2 & 0 & -1 & 0 & -1 & 0 & 0 \\
0 & 1 & -1 & 0 & 0 & 0 & 0 \\
-1 & -1 & 3 & -1 & 0 & 0 & 0 \\
0 & 0 & -1 & 4 & -1 & -1 & -1 \\
-1 & 0 & 0 & -1 & 3 & 0 & -1 \\
0 & 0 & 0 & -1 & 0 & 2 & -1 \\
0 & 0 & 0 & -1 & -1 & -1 & 3
\end{bmatrix}.
$$

continued

Examples 7.1 *continued*

(ii) The complete network, K_n has Laplacian $nI - E$. For the null network, $L = O$. The Laplacian of the path graph P_{n-1} is an $n \times n$ matrix of the form

$$\begin{bmatrix} 1 & -1 & 0 & \cdots & & 0 \\ -1 & 2 & -1 & \ddots & & \vdots \\ 0 & & \ddots & & & 0 \\ \vdots & \ddots & & -1 & 2 & -1 \\ 0 & \cdots & & 0 & -1 & 1 \end{bmatrix}.$$

7.2 Eigenvalues and eigenvectors of the graph Laplacian

Because of its theoretical connections with the Laplacian operator, the graph Laplacian is a very useful tool for analysing a network. In particular, its spectrum *and* its eigenvectors can reveal significant properties of a network.

Since $L = B^T B = D - A$ is symmetric we know that its spectrum is real. Its Gershgorin discs are of the form $\Delta_i = \{z : |z - k_i| \le k_i\}$, where k_i is the degree of node i. Applying Gershgorin's theorem we find that all the eigenvalues of the graph Laplacian lie in the interval $[0, 2k_{\max}]$ where $k_{\max}$ is the maximal degree of a node in the associated network. Since

$$L\mathbf{e} = (D - A)\mathbf{e} = \operatorname{diag}(A\mathbf{e})\mathbf{e} - A\mathbf{e} = A\mathbf{e} - A\mathbf{e} = 0$$

we know that zero is an eigenvalue of any graph Laplacian. Simple bounds on other eigenvalues of L can be attained easily.

Problem 7.1

Suppose we order the eigenvalues of the graph Laplacian L of a simple network G so that

$$\lambda_1 \ge \lambda_2 \ge \cdots \ge \lambda_{n-1} \ge \lambda_n = 0.$$

Show the following bounds hold.

1. $\lambda_{n-1} > 0 \iff G$ is connected.

2. $k_{\max} \le \lambda_1 \le 2k_{\max}$.

3. λ_1 is bounded above by the maximum of the sum of degrees of adjacent nodes. That is,

$$\lambda_1 \le k_{\max}^{(ii)} = \max_{(i,j)\,|\,a_{ij}=1} k_i + k_j.$$

For point 1, note that $(n-1)I - L$ is a nonnegative matrix whose eigenvalues are equal to $(n-1) - \lambda_i$. By the Perron–Frobenius theorem, the largest positive eigenvalue of this matrix, $(n-1)$ has algebraic multiplicity of 1 unless $(n-1)I - L$ is reducible. This can only happen if A is reducible, too, in which case G has more than one connected component.

We can use Rayleigh quotients to establish the lower bound in 2. Suppose $k_{max} = k_i$. Then $\dfrac{\mathbf{e}_i^T L \mathbf{e}_i}{\mathbf{e}_i^T \mathbf{e}_i} = k_i \leq \lambda_1$, where $\mathbf{e}_i$ is the ith column of the identity matrix. The upper bound comes from considering the right hand boundary point of the Gershgorin discs of L.

A simple way to establish the final result is to use Gershgorin's theorem on the matrix $\mathcal{L} = D^{-1}LD$. The diagonal elements of this matrix are the same as those of L and the sum of the moduli of the off diagonal elements in the ith row is

$$\frac{1}{k_i} \sum_{j=1}^{n} k_j a_{ij}. \tag{7.1}$$

There are k_i nonzero terms in the sum and they are bounded above by the maximum degree of the nodes adjacent to i (call this $k_{max}^{(i)}$). Thus the point in the ith Gershgorin disc of maximum size is $k_i + k_{max}^{(i)}$. Taking the maximum over all i gives the desired bound.

Note that (7.1) gives the mean degree of nodes adjacent to i. Thus we can improve our upper bound to

$$\lambda_1 \leq \max_i k_i + m_i,$$

where m_i is the mean degree of nodes adjacent to node i.

The spectra of the adjacency matrix and the graph Laplacian share some relations, but usually we can infer different information from each.

Examples 7.2

(i) For the network in Figure 7.1(a), $\lambda_1 \leq 2k_{max} = k_{max}^{(ij)} = 6$. Each node of degree 3 is connected to two nodes of degree 3 and a node of degree 1. Thus the mean degree of nodes adjacent to a node of degree 3 is (in all cases) 2.333 which improves the upper bound on λ_1 to 5.333. In fact, the largest eigenvalue of the Laplacian is 5.236.

For the network in Figure 7.1(b), $2k_{max} = 8$, $k_{max}^{(ij)} = 7$, and $\max_i k_i + m_i = 6.75$. In this case, $\lambda_1 = 5.449$.

(ii) Recall that G is a k-regular network if all its nodes have degree k. If this is the case, $L = kI - A$ and $\lambda_i \in \sigma(L) \iff k - \lambda_i \in \sigma(A)$. The eigenvectors of L and A are identical for regular networks.

continued

Examples 7.2 *continued*

(iii) K_n is $n - 1$ regular and C_n is 2-regular. Since we know the spectra of their adjacency matrices, we can write down the spectra of their Laplacians, too.

The characteristic polynomial of the Laplacian of K_n is $\det(\lambda I - L) = \lambda(\lambda - n)^{n-1}$ and the eigenvalues of the Laplacian of C_n are

$$\left\{ 2 - 2\cos\left(\frac{2\pi j}{n}\right), j = 1, \ldots, n \right\}$$

or, using standard trigonometric identities,

$$\left\{ 4\sin^2\left(\frac{\pi j}{n}\right), j = 1, \ldots, n \right\}.$$

(iv) As with the adjacency matrix, we can use the eigenvalues and eigenvectors of the Laplacian of the cycle graph C_{2n} to find the spectrum of the path graph (although the details differ). First observe that grouping together pairs of entries in C_{2n} gives us something that looks very like P_{n-1} (see Figure 7.2.[1]).

Given a repeated eigenvalue $\lambda = 2 - \omega - \omega^{-1}$ ($\omega^{2n} = 1$) of the Laplacian, L_C, of C_{2n} we can form a linear combination of the eigenvectors we gave in Chapter 6 to give another eigenvector. In particular,

$$\mathbf{z} = \omega \begin{bmatrix} 1 \\ \omega \\ \omega^2 \\ \vdots \\ \omega^{2n-1} \end{bmatrix} + \begin{bmatrix} 1 \\ \omega^{2n-1} \\ \omega^{2n-2} \\ \vdots \\ \omega \end{bmatrix} = \begin{bmatrix} \omega + 1 \\ \omega^2 + \omega^{2n-1} \\ \vdots \\ \omega^{2n-1} + \omega^2 \\ 1 + \omega \end{bmatrix},$$

a vector whose ith component is the same as its $(2n + 1 - i)$th.

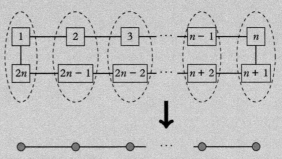

Figure 7.2 *Pairing the nodes in the cycle graph C_{2n} gives the path graph P_{n-1}*

[1] Technically, we are treating P_{n-1} as a *quotient graph* of C_{2n}. The interested reader can find many more details in a text book devoted to graph theory.

Now we show that the first n components of $\mathbf{z}$ constitute an eigenvector of the Laplacian L_P of P_{n-1}.

$$L_P \begin{bmatrix} z_1 \\ z_2 \\ \vdots \\ z_{n-1} \\ z_n \end{bmatrix} = \begin{bmatrix} z_1 - z_2 \\ -z_1 + 2z_2 - z_3 \\ \vdots \\ -z_{n-2} + 2z_{n-1} - z_n \\ -z_{n-1} + z_n \end{bmatrix} = \begin{bmatrix} 2z_1 - z_2 - z_{2n} \\ -z_1 + 2z_2 - z_3 \\ \vdots \\ -z_{n-2} + 2z_{n-1} - z_n \\ -z_{n-1} + 2z_n - z_{n+1} \end{bmatrix} = \lambda \begin{bmatrix} z_1 \\ z_2 \\ \vdots \\ z_{n-1} \\ z_n \end{bmatrix},$$

since $z_1 = z_{2n}$, $z_n = z_{n+1}$ and the action of L_P on the first n components of $\mathbf{z}$ mimics that of L_C.

We conclude that each repeated eigenvalue of C_{2n} is an eigenvalue of P_{n-1}. The remaining eigenvalue is, of course, 0 and so the spectrum of the Laplacian of P_{n-1} is

$$\left\{ 4\sin^2 \left(\frac{\pi j}{2n} \right), j = 0, \ldots, n-1 \right\}.$$

(v) $K_{m,n}$ has Laplacian

$$L = \begin{bmatrix} nI & -E_{mn} \\ -E_{mn}^T & mI \end{bmatrix},$$

where E_{mn} is an $m \times n$ matrix of all ones. We can find its spectrum with a similar technique to that used to find the spectrum of its adjacency matrix.

Let $\begin{bmatrix} \mathbf{x} \\ \mathbf{y} \end{bmatrix}$ be an eigenvector of the eigenvalue λ. Then $L \begin{bmatrix} \mathbf{x} \\ \mathbf{y} \end{bmatrix} = \lambda \begin{bmatrix} \mathbf{x} \\ \mathbf{y} \end{bmatrix}$ gives

$$n\mathbf{x} - E_{mn}\mathbf{y} = \lambda\mathbf{x} \text{ and } m\mathbf{y} - E_{mn}^T\mathbf{x} = \lambda\mathbf{y},$$

so,

$$(m-\lambda)(n-\lambda)\mathbf{x} = (m-\lambda)E_{mn}\mathbf{y} = E_{mn}E_{mn}^T\mathbf{x} = nE_m\mathbf{x}.$$

This means that $(m-\lambda)(n-\lambda)$ is an eigenvalue of nE_m, which in turn means $(m-\lambda)(n-\lambda) = 0$ or $(m-\lambda)(n-\lambda) = mn$. We conclude that the eigenvalues of L are 0, m, n and $m+n$. Taking into account algebraic multiplicities we can establish that the characteristic polynomial in this case is

$$\det(\lambda I - L) = \lambda(\lambda - m - n)(\lambda - m)^{n-1}(\lambda - n)^{m-1}.$$

Notice that the largest eigenvalue of the Laplacian of K_{mn} is $m + n$, confirming the upper bound in point 3 of Problem 7.1 can be attained. It can be shown that complete bipartite networks are the only ones for which the bound can be attained but we omit the details.

As with the adjacency matrix, the spectrum of L can be used to calculate certain useful network quantities. The most widely quoted example is that the number of spanning trees in a network is given by $\lambda_1 \lambda_2 \cdots \lambda_{n-1}$. Notice that if G is not connected then $\lambda_{n-1} = 0$, confirming that there are no spanning trees in a disconnected network.

The eigenvalues of the Laplacian are another example of a graph invariant. For applications where invariance is a key factor, some authors prefer to use the *normalized Laplacian* which can be defined as $\mathcal{L} = D^{-1/2}LD^{-1/2}$ (if a node is isolated we 'fix' the undefined entry in $D^{-1/2}$ by setting it to 1). It can be shown that the spectrum of $\mathcal{L}$ lies in the interval $[0, 2]$ and that for bipartite networks the spectrum is symmetric around 1. For the applications that our predominant interest in this book, the 'ordinary' Laplacian works perfectly well.

7.2.1   The Fiedler vector

A network is connected if the second smallest eigenvalue of its Laplacian is nonzero. On the other hand, if G has k components then the Laplacian has k zero eigenvalues and an orthogonal basis for the eigenvectors associated with the zero eigenvalue can be constructed as follows: for each component form a vector whose ith element is one if and only if node i belongs to that component. Otherwise the ith element is left as zero.

These results are trivial to establish but an important observation is that the eigenvectors of the Laplacian contain useful information for establishing key structures in a network. We have previously considered the notion of edge and vertex connectivity. The spectrum of the graph Laplacian motivates the definition of the *algebraic connectivity* of a network, $a(G)$, which is assigned the value λ_{n-1}. The eigenvector associated with λ_{n-1} has a very important property.

Theorem 7.1 (Fiedler, 1975)   *Suppose $G = (V, E)$ is a connected network with graph Laplacian L whose second smallest eigenvalue is $\lambda_{n-1} > 0$. Let $\mathbf{x}$ be the eigenvector associated with λ_{n-1}. Let $r \in \mathbb{R}$ and partition the nodes in V into two sets*

$$V_1 = \{i \in V \,|\, x_i \geq r\}, \quad V_2 = \{i \in V \,|\, x_i < r\}.$$

Then the subgraphs of G induced by the sets V_1 and V_2 are connected.

The significance of this result is that it gives us a method of *partitioning* a network in a systematic way and ensuring the partitions are still connected. Each of these partitions can be further divided and we may be able to identify key *clusters* of the network. Choosing r to be the median value of $\mathbf{x}$ ensures that the clusters are evenly sized. Another popular choice of r is 0. *Spectral clustering* is one of many ways to divide a network into pieces. The vector $\mathbf{x}$ is known as the *Fiedler vector*.

We will consider a variety of methods for partitioning a network in Chapter 21. Some of these will exploit spectral information but there are a variety of other techniques, too.

Example 7.3

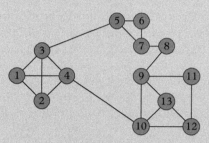

Figure 7.3 *A network with some clusters. The Fiedler vector can reveal them*

Figure 7.3 shows a network we saw in Chapter 2. The nodes have been labelled to highlight the clustering given by the Fiedler vector.

The first five elements of the Fiedler vector have the opposite sign to the others, suggesting one particular partition. Assuming $x_1 > 0$, we find that

$$x_1 = x_2 > x_3 > x_4 > x_5 > x_6 > x_7 > x_{10} > x_8 > x_{13} > x_9 > x_{12} > x_{11}.$$

Grouping nodes according to this ordering highlights several obvious clusters.

FURTHER READING

Chung, F., *Spectral Graph Theory*, American Mathematical Society, 1997.
Fiedler, M., A property of eigenvectors of nonnegative symmetric matrices and its application to graph theory, Czechoslovak Mathematical Journal, 25:619–633, 1975.
Godsil, C. and Royle, G., *Algebraic Graph Theory*, Springer, 2001. Chapter 13.

<table>
<tr><td>

8

</td><td>

Classical Physics Analogies

</td></tr>
</table>

In this chapter

We introduce basic concepts from classical mechanics, such as the Lagrangian and the Hamiltonian of a system, to describe the motion of a particle in space. We use the analogy of a network as a classical mechanics mass–spring system, which allows us to interpret the Laplacian matrix of a network (and its spectrum) physically. We also picture networks as electrical circuits. In this case, we show the effective resistance between a pair of nodes in an electrical circuit is a Euclidean distance between the corresponding nodes in a network. This chapter assumes a basic working knowledge of classical physics.

8.1 Motivation

Analogies and metaphors are very useful in any branch of science. Complex networks are already an abstraction of the connectivity patterns observed in real-world complex systems in which we reduce complex entities to single nodes and their complex relationships to the links of the network. It is very natural, therefore, to use some physical analogies to study these networks so that we can use familiar physical and mathematical concepts to understand the structure and dynamics of networks; and we can use physical intuition about the abstract mathematical concepts to study the structure/dynamics of networks.

In this chapter we focus on analogies based on classical physics. The first corresponds to the use of mass–spring systems in classical mechanics and the second uses electrical circuits. In both cases we show how to gain intuition into the analysis of networks as well as how to import techniques from physics, such as the resistance distance, which can be useful for the study of networks.

For instance, we understand mathematically the Fiedler vector of the Laplacian matrix. Now if we observe that it is analogous to a vibrational mode of the nodes of a mass–spring network, we can develop a physical picture of how this eigenvector splits the nodes of the network into two clusters: one corresponding to the nodes vibrating in one direction and the other to the nodes vibrating in the opposite direction for a certain natural frequency. We fill in the details in Section 8.2.

8.2 Classical mechanical analogies

A principal goal of classical mechanics is to determine the position of a particle as a function of time i.e. the particle trajectory. We will consider finite systems; the simplest of these is a single point with mass m moving in space. In order to compute the trajectory of this particle, $x(t)$, we need to determine the forces $F(x)$ acting on it using Newton's equation

$$F(x) = m\ddot{x}(t), \tag{8.1}$$

where $\ddot{x}(t)$ is the second derivative of the position with respect to time i.e. the acceleration of the particle.

The state of a system with n degrees of freedom is fully determined by n coordinates $x_i(t)$ and n velocities $\dot{x}_i(t)$ for $i = 1, \ldots, n$, and the system is described by the Lagrangian $\mathcal{L}(x, \dot{x}, t)$. The Lagrangian function for a dynamical system is its kinetic energy minus the potential function from which the generalized force components are determined. That is, $\mathcal{L} = T - V$, where T is the kinetic and V the potential energies of the system. For a system with n particles

$$T = \frac{1}{2} \sum_i m_i \dot{x}_i^2, \tag{8.2}$$

and

$$V = \frac{1}{2} \sum_{j,i} k_{ij}(x_j - x_i)^2, \tag{8.3}$$

where the last summation is over all pairs of particles interacting with each other. Thus,

$$\mathcal{L} = \frac{1}{2}\left[\sum_i m_i \dot{x}_i^2 - \sum_{j,i} k_{ij}(x_j - x_i)^2 \right] \tag{8.4}$$

and the momenta of the particle i at time t is

$$p_i(t) = \frac{\partial \mathcal{L}(x, \dot{x}, t)}{\partial \dot{x}_i}. \tag{8.5}$$

The Hamiltonian transformation of the Lagrangian is given by the function

$$\mathcal{H} = \dot{x}_i \frac{\partial \mathcal{L}(x, \dot{x}, t)}{\partial \dot{x}_i} - \mathcal{L}(x, \dot{x}, t) = \sum_{i=1}^{n} \dot{x}_i p_i - \mathcal{L}(x, \dot{x}, t). \tag{8.6}$$

The Hamiltonian function (which we will simply refer to as the Hamiltonian) for a given system is written as

$$\mathcal{H} = T + V. \tag{8.7}$$

The so-called *phase space* of the system $\Gamma\{(x, p)\}$ is formed by the $2n$-tuples $(x, p) = (x_1, x_2, \ldots, x_n, p_1, p_2, \ldots, p_n)$ in which a path of the particle system is determined by the Hamilton equations

$$\dot{x}_i(t) = \frac{\partial \mathcal{H}(x, p, t)}{\partial p_i} = \{x_i, \mathcal{H}\}, \quad \dot{p}_i(t) = -\frac{\partial \mathcal{H}(x, p, t)}{\partial x_i} = \{p_i, \mathcal{H}\},$$

where $\{A, B\}$ is the so-called Poisson bracket in a d-dimensional space, namely,

$$\{A, B\} = \sum_{i=1}^{d} \left(\frac{\partial A}{\partial x_i} \frac{\partial B}{\partial p_i} - \frac{\partial B}{\partial p_i} \frac{\partial A}{\partial x_i} \right). \tag{8.8}$$

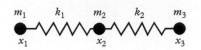

Figure 8.1 *A mass–spring system with masses m_i, spring constants k_i and positions x_i*

We are interested in systems with several interconnected particles. So, for the purpose of illustrating the connections between classical mechanics and network theory, consider a mass–spring system like the system of three masses and two springs illustrated in Figure 8.1.

The kinetic and potential energy for this system can be written as

$$T = \frac{1}{2} m_1 \dot{x}_1^2 + \frac{1}{2} m_2 \dot{x}_2^2 + \frac{1}{2} m_3 \dot{x}_3^2, \quad V = \frac{1}{2} k_1 (x_1 - x_2)^2 + \frac{1}{2} k_2 (x_3 - x_2)^2,$$

and the Lagrangian of the system is given by

$$\mathcal{L} = \frac{1}{2} \left[m_1 \dot{x}_1^2 + m_2 \dot{x}_2^2 + m_3 \dot{x}_3^2 - k_1 (x_1 - x_2)^2 - k_2 (x_3 - x_2)^2 \right]. \tag{8.9}$$

Now if we use the Euler–Lagrange equations

$$\frac{d}{dt} \frac{\partial \mathcal{L}}{\partial \dot{x}_i} - \frac{\partial \mathcal{L}}{\partial x_i} = 0, \tag{8.10}$$

for the system under study we obtain

$$\frac{\partial \mathcal{L}}{\partial \dot{x}_1} = m_1 \dot{x}_1, \quad \frac{\partial \mathcal{L}}{\partial \dot{x}_2} = m_2 \dot{x}_2, \quad \frac{\partial \mathcal{L}}{\partial \dot{x}_3} = m_3 \dot{x}_3, \tag{8.11}$$

and

$$\frac{\partial \mathcal{L}}{\partial x_1} = -k_1 (x_1 - x_2), \quad \frac{\partial \mathcal{L}}{\partial x_2} = k_1 (x_1 - x_2) + k_2 (x_3 - x_2), \quad \frac{\partial \mathcal{L}}{\partial x_3} = -k_2 (x_3 - x_2). \tag{8.12}$$

It is evident that $\dfrac{d}{dt} \dfrac{\partial \mathcal{L}}{\partial \dot{x}_i} = m_i \ddot{x}_i$, which is just F_i according to Newton's equation. Hence, for the system of three masses and two springs,

$$\begin{aligned} m_1 \ddot{x}_1 &= -k_1 x_1 &+ k_1 x_2, \\ m_2 \ddot{x}_2 &= k_1 x_1 &+ (-k_1 - k_2) x_2 &+ k_2 x_3, \\ m_3 \ddot{x}_3 &= &k_2 x_2 &+ -k_2 x_3. \end{aligned} \tag{8.13}$$

which can be written in the matrix-vector form

$$
\begin{bmatrix} m_1 & 0 & 0 \\ 0 & m_2 & 0 \\ 0 & 0 & m_3 \end{bmatrix}
\begin{bmatrix} \ddot{x}_1 \\ \ddot{x}_2 \\ \ddot{x}_3 \end{bmatrix}
=
\begin{bmatrix} -k_1 & k_1 & 0 \\ k_1 & -(k_1 + k_2) & k_2 \\ 0 & k_2 & -k_2 \end{bmatrix}
\begin{bmatrix} x_1 \\ x_2 \\ x_3 \end{bmatrix},
\tag{8.14}
$$

or more concisely as

$$
\ddot{\mathbf{x}}(t) = M^{-1} L \mathbf{x}.
\tag{8.15}
$$

Notice that the matrix on the right-hand side of the equation is just the Laplacian matrix for a weighted network having three nodes and two edges i.e. a path of length 3. If we consider a system with $m_1 = m_2 = m$ and $k_1 = k_2 = k$, then

$$
\begin{bmatrix} \ddot{x}_1 \\ \ddot{x}_2 \\ \ddot{x}_3 \end{bmatrix}
=
\begin{bmatrix} -k/m & k & 0 \\ k & -2k/m & k \\ 0 & k & -k/m \end{bmatrix}
\begin{bmatrix} x_1 \\ x_2 \\ x_3 \end{bmatrix}.
\tag{8.16}
$$

If we now take $m = k = 1$, the characteristic equation

$$
\begin{vmatrix} -1 - \mu & 1 & 0 \\ 1 & -2 - \mu & 1 \\ 0 & 1 & -1 - \mu \end{vmatrix} = 0
\tag{8.17}
$$

has solutions $\mu_1 = 0, \mu_2 = -1, \mu_3 = -3$: the negatives of the eigenvalues of the Laplacian matrix of the corresponding network. In general, (8.16) has solution

$$
\mathbf{x} = (\alpha_1 + \gamma_1 t) \begin{bmatrix} 1 \\ 1 \\ 1 \end{bmatrix} + (\alpha_2 \cos t + \gamma_2 \sin t) \begin{bmatrix} 1 \\ 0 \\ -1 \end{bmatrix}
$$

$$
+ (\alpha_3 \cos \sqrt{3}t + \gamma_3 \sin \sqrt{3}t) \begin{bmatrix} 1 \\ -2 \\ 1 \end{bmatrix}.
\tag{8.18}
$$

This solution is expressed in terms of the eigenvectors associated with the eigenvalues μ_j of the Laplacian matrix.

Our conclusion is that we can interpret the eigenvalues and eigenvectors of the Laplacian matrix of a network in terms of the vibrations of a mass-spring network. The eigenvalues of the Laplacian are the squares of the natural frequencies of the system. The natural frequencies are the frequencies at which a system tends to oscillate in the absence of any external force. On the other hand, the eigenvectors of the Laplacian matrix represent the displacements of each node due to harmonic oscillations. For instance, for the three-node path we have a mode corresponding to the translational movement of the network, i.e. the eigenvector corresponding to $\mu_1 = 0$, plus two additional modes illustrated in Figure 8.2.

Figure 8.2 *Schematic representation of the vibrational modes in P$_2$. The size of the arrows is proportional to the magnitude of the mode*

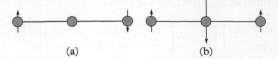

(a)   (b)

This analogy can be very helpful, not least when we use the eigenvectors of the Laplacian to make partitions of the nodes in a network. Recall that the Fiedler vector partitions a network into two clusters and we can think of this partition as the grouping of the nodes according to the vibrational mode of the nodes in the network corresponding to the slowest or fundamental natural frequency of the system, $\mu_2^{1/2}$.

Problem 8.1
Write down the Hamiltonian for the mass-spring network illustrated in Figure 8.3.
Clearly,

$$T = \frac{1}{2}\left(\frac{p_1^2}{m_1} + \frac{p_2^2}{m_2} + \frac{p_3^2}{m_3} + \frac{p_4^2}{m_4} + \frac{p_5^2}{m_5}\right)$$

and

$$V = \frac{1}{2}\left[k_1(x_2-x_1)^2 + k_2(x_3-x_2)^2 + k_3(x_4-x_3)^2 + k_4(x_5-x_4)^2\right.$$
$$\left. + k_5(x_5-x_1)^2 + k_6(x_4-x_2)^2\right].$$

Letting

$$B = \begin{bmatrix} 1 & -1 & 0 & 0 & 0 \\ 0 & 1 & -1 & 0 & 0 \\ 0 & 0 & 1 & -1 & 0 \\ 0 & 0 & 0 & 1 & -1 \\ -1 & 0 & 0 & 0 & 1 \\ 0 & -1 & 0 & 1 & 0 \end{bmatrix}, K = \begin{bmatrix} k_1 & 0 & 0 & 0 & 0 & 0 \\ 0 & k_2 & 0 & 0 & 0 & 0 \\ 0 & 0 & k_3 & 0 & 0 & 0 \\ 0 & 0 & 0 & k_4 & 0 & 0 \\ 0 & 0 & 0 & 0 & k_5 & 0 \\ 0 & 0 & 0 & 0 & 0 & k_6 \end{bmatrix}, \mathbf{x} = \begin{bmatrix} x_1 \\ x_2 \\ x_3 \\ x_4 \\ x_5 \end{bmatrix},$$

where B is the node-to-edge incidence matrix, gives $\mathcal{H}(x,p) = \frac{1}{2}[T + \mathbf{x}^T B^T K B\mathbf{x}]$.

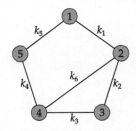

Figure 8.3 *A network with edges labelled by force constants k_i*

8.3 Networks as electrical circuits

An electrical circuit is an arrangement of devices—voltage and current sources, resistors, inductors, and capacitors—which are interconnected in specific ways, forming a network. The connectivity of these devices determines the relation between the physical variables of the system, such as voltages and currents. The circuit has nodes, which represent the junction between two or more wires, and edges which represent the wires. The wires are characterized by having some resistance to the current flow and the nodes by the existence of voltage, which is created by the differences in the currents flowing through the edges of the circuit and their respective resistance. According to *Ohm's law* the voltage is related to the current I and the resistance R according to the relation $V = IR$. The inverse of the resistance is known as the *conductance* of the corresponding edge.

For a circuit we can represent the voltages, currents, and resistances of all the edges by means of vectors. That is, let $\mathbf{v}$ be the vector representing the voltages at each node of a circuit. Then the current at each edge is given by the vector representation of Ohm's law

$$\mathbf{i} = R^{-1}B^T\mathbf{v}, \tag{8.19}$$

where B is the node-to-edge incidence matrix and R is a diagonal matrix of edge resistances. Using the conductance matrix $C = R^{-1}$, Ohm's law is

$$\mathbf{i} = CB^T\mathbf{v}. \tag{8.20}$$

In general, a circuit contains external sources of voltage and current used to provide fixed sources of energy to drive the circuit. There is a relationship between the voltage and current sources known as Norton equivalence. Consequently, we only need to consider the external current sources as any voltage external source can be transformed into its equivalent current one. Furthermore by using *Kirchhoff's current law*, which states that the algebraic sum of the currents for all edges that meet in a common node is zero, we can write

$$\mathbf{i}_{ext} = B^T\mathbf{i}, \tag{8.21}$$

and from (8.20),

$$\mathbf{i}_{ext} = B^T CB^T\mathbf{v} = L\mathbf{v}, \tag{8.22}$$

where L is the Laplacian of a weighted network.

If our aim is to find the voltages at the nodes generated by the external sources of current we need to solve (8.22). That is, we would like to find

$$\mathbf{v} = L^{-1}\mathbf{i}_{ext}. \tag{8.23}$$

However, Laplacians have zero eigenvalues, so L^{-1} does not exist. A general trick used to overcome the singularity in (8.22) is to ground a node, which in practice means that we remove the corresponding row and column of L, and we solve the reduced system $\mathbf{v}_0 = L_0^{-1}\mathbf{i}_{ext_0}$. Another alternative is to let $\mathbf{v} = L^+\mathbf{i}_{ext}$, where

$$L^+ = \left(L + \frac{1}{n}\mathbf{e}\mathbf{e}^T\right)^{-1} - \frac{1}{n}\mathbf{e}\mathbf{e}^T \tag{8.24}$$

is the Moore–Penrose pseudoinverse of the Laplacian matrix.

Let us now suppose that we want to calculate the voltage at each node induced when a current of 1 amp enters at node p and a current of -1 amp leaves node q. Using (8.24) we have

$$\mathbf{v} = \left[(L_{p1}^+ - L_{q1}^+) \quad (L_{p2}^+ - L_{q2}^+) \quad \cdots \quad (L_{pn}^+ - L_{qn}^+) \right]^T. \tag{8.25}$$

If we want the difference in voltage created at the nodes p and q, we only need

$$\Omega_{pq} = v_p - v_q = (L_{pp}^+ - L_{qp}^+) - (L_{pq}^+ - L_{qq}^+) = L_{pp}^+ + L_{qq}^+ - 2L_{pq}^+. \tag{8.26}$$

Notice that $L_{pq}^+ = L_{qp}^+$ because the network is undirected.

The voltage difference Ω_{pq} represents the effective resistance between the two nodes p and q, which indicates the potential drop measured when a unit current is injected at node p and extracted at node q. Hence

$$\Omega_{pq} = (\mathbf{e}_p - \mathbf{e}_q)^T L^+ (\mathbf{e}_p - \mathbf{e}_q). \tag{8.27}$$

Problem 8.2

Show that the effective resistance is a Euclidean distance between a pair of nodes.

By definition, the effective resistance is given by

$$\Omega_{pq} = L_{pp}^+ + L_{qq}^+ - 2L_{pq}^+.$$

It turns out that the spectral decomposition of the Moore–Penrose pseudoinverse of the Laplacian is

$$L_{pq}^+ = \sum_{j=2}^{n} \mu_j^{-1} \mathbf{q}_j(p) \mathbf{q}_j(q),$$

where $0 = \mu_1 < \mu_2 \leq \cdots \leq \mu_n$ are the eigenvalues of the Laplacian and $\mathbf{q}_j$ the eigenvector associated with μ_j. So

$$\Omega_{pq} = \sum_{j=2}^{n} \frac{1}{\mu_j} [\mathbf{q}_j(p) - \mathbf{q}_j(q)]^2,$$

which can be expressed in matrix-vector form as

$$\Omega_{pq} = (\mathbf{q}_p - \mathbf{q}_q)^T M^{-1} (\mathbf{q}_p - \mathbf{q}_q),$$

where

$$\mathbf{q}_r = \left[\mathbf{q}_2(r) \quad \mathbf{q}_3(r) \quad \cdots \quad \mathbf{q}_n(r) \right]^T$$

and M is a diagonal matrix of the eigenvalues of the Laplacian with the trivial one replaced by a nonzero. Hence

$$\Omega_{pq} = (M^{-1/2}(\mathbf{q}_p - \mathbf{q}_q))^T M^{-1/2}(\mathbf{q}_p - \mathbf{q}_q)(M^{-1/2}\mathbf{q}_p - M^{-1/2}\mathbf{q}_q)^T$$
$$\times (M^{-1/2}\mathbf{q}_p - M^{-1/2}\mathbf{q}_q).$$

Letting $\mathbf{y}_r = M^{-1/2}\mathbf{q}_r$ allows us to write

$$\Omega_{pq} = (\mathbf{y}_p - \mathbf{y}_q)^T(\mathbf{y}_p - \mathbf{y}_q) = \|\mathbf{y}_p - \mathbf{y}\|^2.$$

Consequently, the effective resistance between two nodes in a network is a squared Euclidean distance. In fact, it is known in the literature as the resistance distance between the corresponding nodes.

The resistance distance between all pairs of nodes in the network can be represented in a matrix form, namely as the resistance matrix Ω. This matrix can be written as

$$\Omega = \mathbf{e}\mathbf{l}^T + \mathbf{l}\mathbf{e}^T - 2(L + (1/n)E)^{-1}, \tag{8.28}$$

where $\mathbf{l}$ is the diagonal of the inverse of $L + (1/n)E$.

Example 8.1

Let us compare the shortest path distance and the resistance distance for the network illustrated in Figure 8.3. These are

$$D = \begin{bmatrix} 0 & 1 & 2 & 2 & 1 \\ & 0 & 1 & 1 & 2 \\ & & 0 & 1 & 2 \\ & & & 0 & 1 \\ & & & & 0 \end{bmatrix} \text{ and } \Omega = \begin{bmatrix} 0 & 0.727 & 1.182 & 0.909 & 0.727 \\ & 0 & 0.636 & 0.545 & 0.909 \\ & & 0 & 0.636 & 1.182 \\ & & & 0 & 0.727 \\ & & & & 0 \end{bmatrix}.$$

Although the pairs $(1,3)$ and $(1,4)$ are the same distance apart according to the shortest path distance, the second pair is closer according to the resistance distance. The reason is that 1 and 4 are part of a square in the network while the smallest cycle 1 and 3 are part of is a pentagon. In fact, the shortest resistance distance is between the nodes 2 and 4, which is the only pair of nodes which are part of a triangle and a square at the same time. Thus, the resistance distance appears to take into account not only the length of the shortest path connecting two nodes, but also the cycles involving them.

FURTHER READING

Doyle, P.G. and Snell, J.L., *Random Walks and Electric Networks*, John Wiley and Sons, 1985.

Estrada, E. and Hatano, N., A vibrational approach to node centrality and vulnerability in complex networks, Physica A **389**:3648–3660, 2010.

Klein, D.J. and Randić, M., Resistance distance, Journal of Mathematical Chemistry **12**:81–95, 1993.

Susskind, L. and Hrabovsky, G., *Classical Mechanics: The Theoretical Minimum*, Penguin, 2014.

Degree Distributions

In this chapter

We start by introducing the concept of degree distribution. We analyse some of the most common degree distributions found in complex networks, such as the Poisson, exponential, and power-law degree distributions. We explore some of the main problems found when fitting real-world data to certain kinds of distributions.

9.1 Motivation

The study of degree distributions is particularly suited to the analysis of complex networks. This kind of statistical analysis of networks is inappropriate for the small graphs typically studied in graph theory. The aim is to find the best fit for the probability distribution of the node degrees in a given network. From a simple inspection of adjacency matrices of networks one can infer that there are important differences in the way degrees are distributed. In this chapter we introduce the tools which allow us to analyse these distributions in more detail.

9.2 General degree distributions

Let $p(k) = n(k)/n$, where $n(k)$ is the number of nodes of degree k in a network of size n. Thus $p(k)$ represents the probability that a node selected uniformly at random has degree k and we can represent the degree distribution of the network by plotting $p(k)$ against k.

What should we expect to find when we look at degree distributions? Consider a network with n nodes that is generated at random so that an edge between two nodes exists with probability p. The average degree will be np and the probability that a particular node v has degree k will follow a binomial distribution

$$\Pr(\deg(v) = k) = \binom{n-1}{k} p^k (1-p)^{n-1-k}.$$

For large n (and $np = \bar{k}$ constant) one finds in the limit that node degree follows a Poisson distribution

$$\Pr(\deg(v) = k) \rightarrow \frac{\bar{k}^k e^{-\bar{k}}}{k!}.$$

As λ grows, this distribution in turn becomes a normal distribution.

Another way we could imagine a network emerging is that we start with a single node and nodes are added one at a time and attach themselves randomly to existing nodes. The newer the node, the lower its expected degree and one finds that a network generated in this way will have an exponential degree distribution

$$\Pr(\deg(v) = k) = Ae^{-k/\bar{k}}.$$

Many real networks can be found that have degree distributions similar to those illustrated. But there is another distribution one frequently sees that deserves attention. That is the power-law distribution

$$\Pr(\deg(v) = k) = Bk^{-\gamma}.$$

One can show that such a distribution emerges in a random graph where as new nodes are added they attach preferentially to nodes with high degree. This is a model of popularity, and is often the predicted behaviour one expects to see in social networks.

If a large network has a power-law degree distribution then one can expect to see many end vertices and other nodes of low degree and a handful of very well connected nodes.

In Figure 9.1 we illustrate some common distributions found in complex networks.

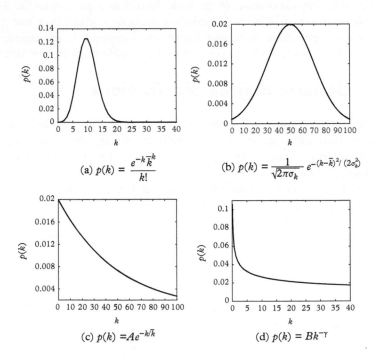

(a) $p(k) = \dfrac{e^{-k}\bar{k}^{k}}{k!}$

(b) $p(k) = \dfrac{1}{\sqrt{2\pi\sigma_k}}\,e^{-(k-\bar{k})^2/(2\sigma_k^2)}$

(c) $p(k) = Ae^{-k/\bar{k}}$

(d) $p(k) = Bk^{-\gamma}$

Figure 9.1 *Common examples of degree distributions found in complex networks*

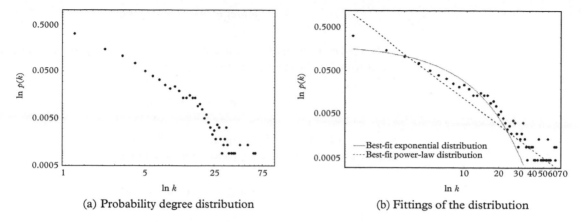

(a) Probability degree distribution   (b) Fittings of the distribution

Figure 9.2 *Degree distribution of a real-world network*

The degree distributions of real-world networks, however, do not look so smooth as the ones illustrated in Figure 9.1 and there are a number of difficulties in fitting the best model to describe the experimental data. For example, the relationship of $p(k)$ and k can be erratic (see Figure 9.2a); there may not be sufficient data to fit a statistically significant model; or there may be many statistical distributions to which the same dataset can be fitted (see Figure 9.2b).

Often the data are most erratic for large k. There are two main approaches that can be used to reduce the noisy effect in the tail of probability distributions. The first is to build a histogram in which the bin sizes increase exponentially with degree. The second is to consider the cumulative distribution function (CDF) defined as

$$P'(k) = \sum_{k'=1}^{k} p(k').$$ (9.1)

This represents the probability of choosing at random a node with degree smaller than or equal to k.

Common CDFs for degree distributions are

$$\text{Poisson: } P'(k) = e^{-\bar{k}} \sum_{i=1}^{\lfloor k \rfloor} \frac{\bar{k}^i}{i}.$$

$$\text{Exponential: } P'(k) = 1 - e^{-k/\bar{k}}.$$

$$\text{Power-law: } P'(k) = 1 - Bk^{-\gamma+1}.$$

9.3 Scale-free networks

Networks with power-law degree distributions are usually referred to as 'scale-free'. If $p(k) = Ak^{-\gamma}$ then scaling the degree by a constant factor c produces a proportionate scaling of the probability distribution

$$p(k, c) = A(ck)^{-\gamma} = c^{-\gamma} p(k).$$

Power-law relationships are usually represented on a logarithmic scale since the distribution then appears as the straight line

$$\ln p(k) = -\gamma \ln k + \ln A,$$

where $-\gamma$ is the slope and $\ln A$ the intercept of the function. Scaling by a constant factor c only alters the intercept and the slope is preserved so

$$\ln p(k, c) = -\gamma \ln k + \ln A - \gamma \ln c$$

The existence of a scaling law in a system means that the phenomenon under study will reproduce itself on different time and/or space scales. That is, it has *self-similarity*.

In network theory we often use the quantity

$$P(k) = \sum_{k' \geq k} p(k') = 1 - P'(k) + p(k). \tag{9.2}$$

in place of the standard CDF. It represents the probability of choosing at random a node with degree larger than or equal to k. For example, if $p(k) = Ak^{-\gamma}$ then

$$P(k) = \int_k^\infty Ak'^{-\gamma} \, dk' = Bk^{-\gamma+1},$$

for some constant B. For a power-law or an exponential distribution ($P(k) = e^{-k/\bar{k}}$) it is easier to use $P(k)$ and in this book we will always use it as the CDF in place of $P'(k)$ unless stated otherwise. In Figure 9.3 we illustrate the CDF for the PDF displayed in Figure 9.2a.

All sorts of power-law degree distributions have been 'observed' in the literature. Some of these distributions have been subsequently reviewed and better fits with other probability distributions have been found in some cases. The more generic term of 'fat-tail' degree distribution has been proposed to encompass a range of distributions which including power-law, log-normal, Burr, log-Gamma, and Pareto.

Problem 9.1

A representation of the internet as an autonomous system gives a network with 4,885 nodes. The CDF of its degree distribution is illustrated in Figure 9.4 on a log–log plot. The slope of the best fitting straight line is -1.20716 and the intercept is -0.11811.

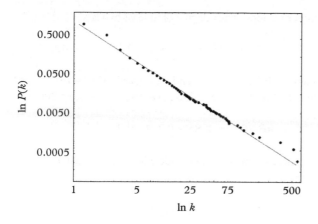

Figure 9.3 *Cumulative degree distribution of the network whose PDF is illustrated in Figure 9.2a*

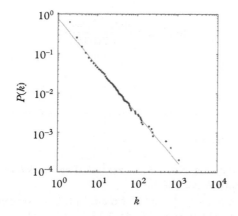

Figure 9.4 *The CDF of an internet's degree distribution*

1. Calculate the percentage of pendant nodes in this network.
2. How many nodes are expected to have degree greater than or equal to 200?
3. Assuming that the network has a unique node of maximal degree, k_{max}, calculate the expected value of k_{max}.

Recalling that the cumulative degree $P(k)$ represents the probability of finding a node with degree larger at least k we note that

$$P(k) = 0.7619k^{-1.20716}. \tag{9.3}$$

The number of nodes with degree at least one is $nP(1)$ (since $P(k) = n(k)/n$, where $n(k)$ is the number of nodes with degree at least k) and

$$nP(1) = 4885 \times 0.7619 = 3722. \tag{9.4}$$

The number of nodes with degree of at least two is

$$nP(2) = 4885 \times 0.7619 \times 2^{-1.20716} = 1612. \tag{9.5}$$

Consequently, the number of pendant nodes is $3722 - 1612 = 2110$ or 43.2%. The number of nodes expected to have degree of at least 200 is

$$nP(200) = 4885 \times 0.7619 \times 200^{-1.20716} = 6. \tag{9.6}$$

Assuming there is only one such node with degree equal to k_{max} then its degree will satisfy the equation $nP(k_{max}) = 1$. So

$$0.7619k_{max}^{-1.20716} = n^{-1}$$

or

$$\log k_{max} = \frac{\log(0.7619n)}{1.20716}.$$

Hence $k_{max} = 908$.

....

FURTHER READING

Caldarelli, G., *Scale-Free Networks: Complex Webs in Nature and Technology*, Oxford University Press, 2007.

Clauset, A., Rohilla Shalizi, C., and Newman, M.E.J., Power-law distributions in empirical data, SIAM Review, 51:661–703, 2010.

Newman, M.E.J., *Networks: An Introduction*, Oxford University Press, 2010, Chapter 8.

Clustering Coefficients of Networks

10

In this chapter

We introduce the notion of the clustering coefficient. We define the Watts–Strogatz and the Newman clustering coefficients, give some of their mathematical properties, and compare them.

10.1 Motivation

Many real-world networks are characterized by the presence of a relatively large number of triangles. This characteristic feature of a network is a general consequence of high transitivity. For instance, in a social network it is highly probable that if Bob and Phil are both friends of Joe then they will eventually be introduced to each other by Joe, closing a transitive relation, i.e. forming a triangle. Our relative measure is between the proportion of triangles existing in a network and the potential number of triangles it can support given the degrees of its nodes. In this chapter we study two methods of quantifying this property of a network, known as its clustering coefficient.

10.2 The Watts–Strogatz clustering coefficient

The first proposal for a clustering coefficient was put forward by Watts and Strogatz in 1998. If we suppose that the clustering of a node is proportional to

$$C_i = \frac{\text{number of transitive relations of node } i}{\text{total number of possible transitive relations of node } i} \tag{10.1}$$

and t_i designates the number of triangles attached to node i of degree k_i then

$$C_i = \frac{t_i}{k_i(k_i-1)/2} = \frac{2t_i}{k_i(k_i-1)}. \tag{10.2}$$

Thus the average clustering coefficient of the network is

$$\overline{C} = \frac{1}{n} \sum_i C_i. \tag{10.3}$$

Example 10.1

Consider the network illustrated in Figure 10.1.

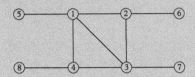

Figure 10.1 *A network used to calculate a clustering coefficient*

Nodes 1 and 3 are equivalent. They both take part in two triangles and their degrees is 4. Thus,

$$C_1 = C_3 = \frac{2 \cdot (2)}{4 \cdot 3} = \frac{1}{3}. \tag{10.4}$$

In a similar way, we obtain

$$C_2 = C_4 = \frac{1}{3}. \tag{10.5}$$

Notice that because nodes 5–8 are not involved in any triangle we have, $C_{i \geq 5} = 0$. Consequently,

$$\overline{C} = \frac{1}{8} \left(\frac{4}{3} \right) = \frac{1}{6}. \tag{10.6}$$

10.3   The Newman clustering coefficient

Another way of quantifying the global clustering of a network is by means of the Newman clustering coefficient, also known as the *transitivity index* of the network. Let $t = |C_3|$ be the total number of triangles, and let $|P_2|$ be the number of paths of length two in the network (representing all potential three-way relationships). Then,

$$C = \frac{3t}{|P_2|} = \frac{3|C_3|}{|P_2|}. \tag{10.7}$$

Example 10.2

Consider again the network illustrated in Figure 10.1. We can obtain the number of triangles in that network by using the spectral properties of the adjacency matrix. That is,

$$t = \frac{1}{6} \operatorname{tr}(A^3) = 2. \tag{10.8}$$

The number of paths of length two in the network can be obtained using the following formula (which we will justify in Chapter 13).

$$|P_2| = \sum_{i=1}^{n} \binom{k_i}{2} = \sum_{i=1}^{n} \frac{k_i(k_i - 1)}{2} = 18. \tag{10.9}$$

Thus,

$$C = \frac{3 \times 2}{18} = \frac{1}{3}. \tag{10.10}$$

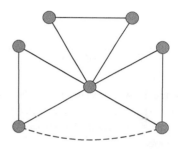

Figure 10.2 *A network formed by triangles joined at a central node. The dashed line indicates the existence of other triangles.*

Problem 10.1

Consider the network illustrated in Figure 10.2. Obtain an expression for the average clustering, $\overline{C}$, and the network transitivity, C, in terms of the number of nodes n. Analyse your results as $n \to \infty$.

To answer this problem, observe that there are two types of nodes in the network, which will be designated as i and j (see Figure 10.3). There is one node of type i and $n - 1$ nodes of type j.

The average clustering coefficient is then,

$$\overline{C} = \frac{C_i + (n-1)C_j}{n}. \tag{10.11}$$

Evidently, $C_j = 1$ and

$$C_i = \frac{2t}{k_i(k_i - 1)}, \tag{10.12}$$

where t is the number of triangles in the network (note that node i is involved in all of them) and k_i is the degree of that node. It is easy to see that $k_i = 2t = n - 1$. Then,

$$C_i = \frac{2t}{2t(2t - 1)} = \frac{1}{2t - 1} = \frac{1}{n - 2} \tag{10.13}$$

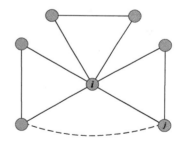

Figure 10.3 *A labelling of Figure 10.2 to indicate nodes with different properties*

and

$$\overline{C} = \frac{C_i + (n-1)C_j}{n} = \frac{\left(\dfrac{1}{n-2}\right) + (n-1)\cdot 1}{n} = \frac{1}{n(n-2)} + \frac{n-1}{n}. \qquad (10.14)$$

Now, for the Newman transitivity coefficient we have

$$|P_2| = \sum_t \binom{k_t}{2} = \frac{1}{2}\sum_t k_t(k_t - 1) = \frac{(n-1)}{2}(2\times 1) + \frac{2t(2t-1)}{2}$$
$$= 2t + t(2t-1) = t(2t+1). \qquad (10.15)$$

Thus,

$$C = \frac{3t}{|P_2|} = \frac{3t}{t(2t+1)} = \frac{3}{2t+1} = \frac{3}{n}. \qquad (10.16)$$

As the number of nodes tends to infinity we have:

$$\lim_{n\to\infty} \overline{C} = \lim_{n\to\infty} \frac{1}{n(n-2)} + \lim_{n\to\infty} \frac{n-1}{n} = 0 + 1 = 1, \qquad (10.17)$$

$$\lim_{n\to\infty} C = \lim_{n\to\infty} \frac{3}{n} = 0. \qquad (10.18)$$

This indicates that the indices are accounting for different structural characteristics of a network.

In general, the Watts–Strogatz index quantifies how clustered a network is locally, while the Newman index indicates how clustered the network is as a whole. Often there is a good correlation between both indices for real-world networks as illustrated in Figure 10.4.

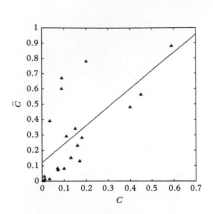

Figure 10.4 *Correlation between Watts–Strogatz ($\overline{C}$) and Newman (C) clustering coefficients for 20 real-world networks*

···

FURTHER READING

Estrada, E., *The Structure of Complex Networks: Theory and Applications*, Oxford University Press, 2011, Chapter 4.5.1.

Newman, M.E.J., *Networks: An Introduction*, Oxford University Press, 2010, Chapter 7.9.

Random Models of Networks

11

In this chapter

We introduce simple and general models for generating random networks: the Erdös–Rényi model, the Barabási–Albert model, and the Watts–Strogatz model. We study some of the general properties of the networks generated by using these models, such as their densities, average path length, and clustering coefficient, as well as some of their spectral properties.

11.1 Motivation

Every time that we look at a real-world network and analyse its most important topological properties it is worth considering how that network was created. In other words, we have to figure out what are the mechanisms behind the evolution of a group of nodes and links which give rise to the topological structure we observe.

Intuitively we can think about a model in which pairs of nodes are connected with some probability. That is, if we start with a collection of n nodes and for each of the $n(n-1)/2$ possible links, we connect a pair of nodes u, v with certain probability $p_{u,v}$. Then, if we consider a set of network parameters to be fixed and allow the links to be created by a random process, we can create models that permit us to understand the influence of these parameters on the structure of networks. Here we study some of the better known models that employ such mechanisms.

11.2 The Erdös–Rényi model of random networks

In this model, put forward by Erdös and Rényi in 1959, we start with n isolated nodes. We then pick a pair of nodes and with probability $p > 0$ we add a link between them. In practice we fix a parameter value p from which we generate the network. For each pair of nodes we generate a random number, r, uniformly from

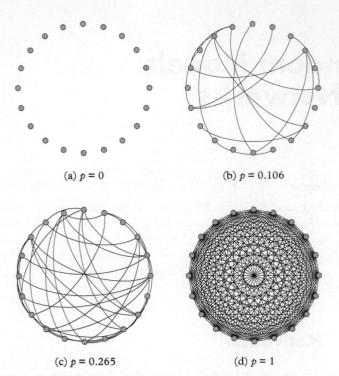

(a) $p = 0$　　　　　　(b) $p = 0.106$

(c) $p = 0.265$　　　　　(d) $p = 1$

Figure 11.1 *Erdös–Rényi random networks for different probabilities p*

[0, 1] and if $p > r$ we add a link between them. Consequently, if we select $p = 0$ the network will remain fully disconnected forever and if $p = 1$ we end up with a complete graph. In Figure 11.1 we illustrate some examples of Erdös–Rényi random networks with 20 nodes and different linking probabilities.

The Erdös–Rényi (ER) random network is written as either $G_{ER}(n, m)$ or $G_{ER}(n, p)$. A few properties of the random networks generated by this model are summarized below.

1.　The expected number of edges is $\bar{m} = \dfrac{n(n-1)p}{2}$.

2.　The expected node degree is $\bar{k} = (n-1)p$.

3.　The degrees follow a Poisson distribution $p(k) = \dfrac{e^{-\bar{k}}\bar{k}^k}{k}$ as illustrated in Figure 11.2 for ER random networks with 1,000 nodes and 4,000 links. The solid line is the expected distribution and the dots represent the values for the average of 100 realizations.

4.　The average path length for large n is

$$\bar{l}(G) = \frac{\ln n - \gamma}{\ln(pn)} + \frac{1}{2}, \qquad (11.1)$$

where $\gamma \approx 0.577$ is the Euler–Mascheroni constant.

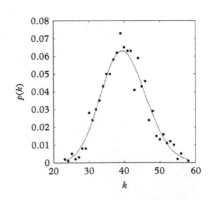

Figure 11.2 *Empirical Erdös–Rényi degree distribution*

5. The average clustering coefficient is $\overline{C} = p$.

6. As p increases, most nodes tend to be clustered in one giant component, while the rest of nodes are isolated in very small components. In Figure 11.3 we illustrate the change of the size of the main connected component in an ER random network with 1,000 nodes, as a function of the linking probability.

7. The structure of $G_{ER}(n, p)$ changes as a function of $p = \overline{k}/(n-1)$, giving rise to the following three stages.

   (a) *Subcritical* $\overline{k} < 1$, where all components are simple and very small. The size of the largest component is $S = O(\ln n)$.

   (b) *Critical* $\overline{k} = 1$, where the size of the largest component is $S = O(n^{2/3})$.

   (c) *Supercritical* $\overline{k} > 1$, where the probability that $(f-\varepsilon)n < S < (f+\varepsilon)n$ is 1 when $n \to \infty$, $\varepsilon > 0$, and where $f = f(\overline{k})$ is the positive solution of the equation $e^{-\overline{k}f} = 1 - f$. The rest of the components are very small, with the second largest having size about $\ln n$.

In Figure 11.4 we illustrate this behaviour for an ER random network with 100 nodes and different linking probabilities. The nodes in the largest connected component are drawn in a darker shade.

8. The largest eigenvalue of the adjacency matrix in an ER network grows proportionally to n so that $\lim\limits_{n \to \infty} \dfrac{\lambda_1(A)}{n} = p$.

9. The second largest eigenvalue grows more slowly than λ_1. In fact,

$$\lim_{n \to \infty} \frac{\lambda_2(A)}{n^\varepsilon} = 0$$

for every $\varepsilon > 0.5$.

   We will make the simplifying assumption $\lambda_2 \to 0$ as $n \to \infty$ in some of our examples. See the appendix for more detail.

10. The most negative eigenvalue grows in a similar way to $\lambda_2(A)$. Namely,

$$\lim_{n \to \infty} \frac{\lambda_n(A)}{n^\varepsilon} = 0$$

for every $\varepsilon > 0.5$.

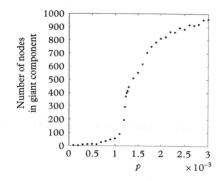

Figure 11.3 *Connectivity of Erdös–Rényi random networks*

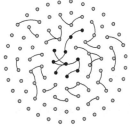 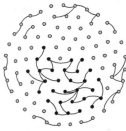 

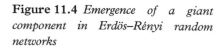

(a) $p = 0.0075$, $S = 10$ (b) $p = 0.01$, $S = 30$ (c) $p = 0.025$, $S = 92$

Figure 11.4 *Emergence of a giant component in Erdös–Rényi random networks*

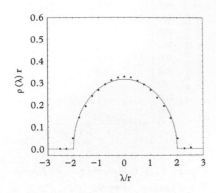

Figure 11.5 *Spectral density for a network generated with the ER model*

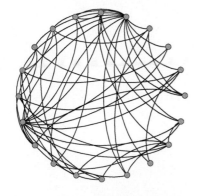

Figure 11.6 *A Barabási–Albert network with n = 20 and m = 4*

11. The spectral density of an ER random network follows Wigner's semicircle law. That is, almost all of the eigenvalues of an ER random network lie in the range $[-2r, 2r]$ where $r = \sqrt{np(1-p)}$ and within this range the density function is given by

$$\rho(\lambda) = \frac{\sqrt{4 - \lambda^2}}{2\pi}.$$

This is illustrated in Figure 11.5.

11.3   The Barabási–Albert model

The ER model generates networks with Poisson degree distributions. However, it has been empirically observed that many networks in the real-world have a fat-tailed degree distribution of some kind, which varies greatly from the distribution observed for ER random networks. A simple model to generate networks in which the probability of finding a node of degree k decays as a power law of the degree was put forward by Barabási and Albert in 1999. We initialize with a small network with m_0 nodes. At each step we add a new node u to the network and connect it to $m \leq m_0$ of the existing nodes $v \in V$. The probability of attaching node u to node v is proportional to the degree of v. That is, we are more likely to attach new nodes to existing nodes with high degree. This process is known as preferential attachment.

We can assume that our initial random network is connected and of ER type with m_0 nodes, $G_{ER} = (V, E)$. In this case the Barabási–Albert (BA) algorithm can be understood as a process in which small inhomogeneities in the degree distribution of the ER network grow in time. A typical BA network is illustrated in Figure 11.6.

Networks generated by this model have several global properties

1. The probability that a node has degree $k \geq d$ is given by

$$p(k) = \frac{2d(d-1)}{k(k+1)(k+2)} \approx k^{-3}. \tag{11.2}$$

That is, the distribution is close to a power law as illustrated in Figure 11.7.

2. The cumulative degree distribution is $P(k) \approx k^{-2}$, illustrated on a log–log scale in Figure 11.8

3. The expected value for the clustering coefficient, $\overline{C}$, approximates $\frac{d-1}{8} \frac{\log^2 n}{n}$ as $n \to \infty$.

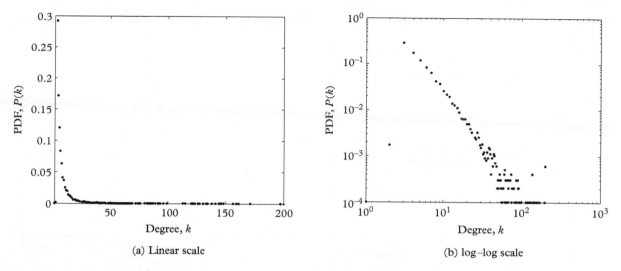

(a) Linear scale   (b) log–log scale

Figure 11.7 *The characteristic power-law degree distribution of a BA network*

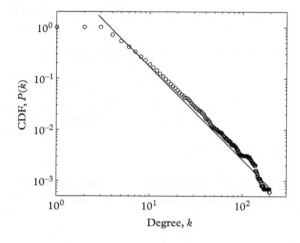

Figure 11.8 *Cumulative degree distribution for a network generated with the BA model*

4. The average path length is given by

$$\bar{l} = \frac{\ln n - \ln(d/2) - 1 - \gamma}{\ln \ln n + \ln(d/2)} + \frac{3}{2}, \qquad (11.3)$$

where again γ is the Euler–Mascheroni constant. Comparing (11.3) with (11.1) we find that for the same number of nodes and average degree, BA networks have smaller average path length than their ER analogues. We illustrate this in Figure 11.9a, which shows the change in the average path length of random networks created with the BA and ER models as the number of nodes increases.

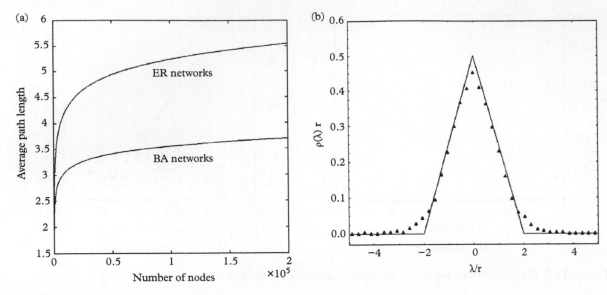

Figure 11.9 *(a) Comparison of the small-worldness of BA and ER networks. (b) Spectral density of a model BA network*

5. The density of eigenvalues follows a triangle distribution

$$\rho(\lambda) = \begin{cases} (\lambda + 2)/4, & -2 \leq \lambda/r \leq 0, \\ (2 - \lambda)/4, & 0 \leq \lambda/r \leq 2, \\ 0, & \text{otherwise.} \end{cases} \tag{11.4}$$

This is illustrated in Figure 11.9b.

The BA model can been generalized to fit general power-law distributions where the probability of finding a node with degree k decays as a negative power of the degree: $p(k) \sim k^{-\gamma}$.

11.4   The Watts–Strogatz model

The phrase 'six degrees of separation' is commonly used to express how surprisingly closely connected we are to each other in terms of shared acquaintances. The phrase originates from a famous experiment in network theory. Stanley Milgram carried out the experiment in 1967. He asked some randomly selected people in the US cities of Omaha (Nebraska) and Wichita (Kansas) to send a letter to a target person who lived in Boston (Massachusetts) on the East Coast. The rules stipulated that the letter should be sent to somebody the sender knew personally. Although the senders and the target were separated by about 2,000 km and there were 200 million inhabitants in the USA at the time, Milgram found two characteristic effects. First, the average number of steps needed for the letters to arrive to its target was around six. And second, there was a large group inbreeding, which resulted in acquaintances of one individual feeding a letter back into his/her own circle, thus usually eliminating new contacts.

Although the ER model reproduces the first characteristic very well, i.e. that most nodes are separated by a very small average path length, it fails in reproducing the second. That is, the clustering coefficient in the ER network is very small in comparison with those observed in real-world systems. The model put forward by Watts and Strogatz in 1998 tries to sort out this situation.

First we form the circulant network with n nodes connected to k neighbours. We then rewire some of its links: each of the original links has a probability p (fixed beforehand) of having one of its end points moved to a new randomly chosen node. If p is too high, meaning almost all links are random, we approach the ER model.

The general process is illustrated in Figure 11.10. On the left is a circulant graph and on the right is a random ER network. Somewhere in the middle are the so-called 'small-world' networks.

In Figure 11.11 we illustrate the rewiring process, which is the basis of the Watts–Strogatz (WS) model for small-world networks. Starting from a regular circulant network with $n = 20, k = 6$ links are rewired with different choices of probability p.

Networks generated by the WS model have several general properties, listed below.

1. The average clustering coefficient is given by $\overline{C} = \dfrac{3(k-2)}{4(k-1)}$. For large values of k, $\overline{C}$ approaches 0.75.

2. The average path length decays very fast from that of a circulant graph,

$$\bar{l} = \frac{(n-1)(n+k-1)}{2kn}, \qquad (11.5)$$

to approach that of a random network. In Figure 11.12 we illustrate the effect of changing the rewiring probability on both the average path length

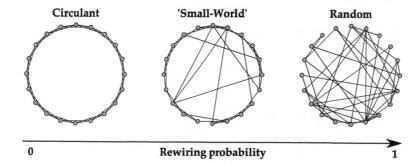

| Circulant | 'Small-World' | Random |

0　　　　　　　　　　**Rewiring probability**　　　　　　　　1

Figure 11.10 *Schematic representation of the Watts–Strogatz rewiring process*

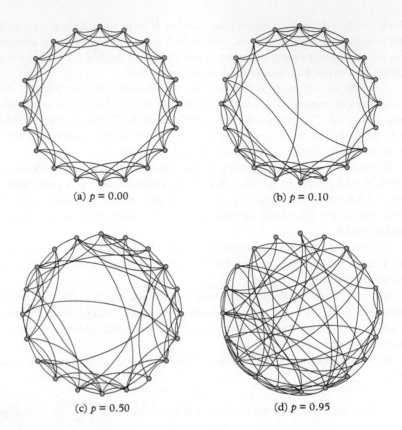

Figure 11.11 *Watts–Strogatz random networks for different rewiring probabilities p*

(a) p = 0.00   (b) p = 0.10

(c) p = 0.50   (d) p = 0.95

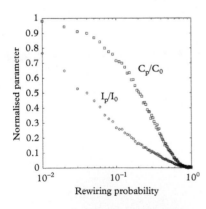

Figure 11.12 *Changes in network statistics for Watts–Strogatz graphs*

and clustering coefficient on random graphs generated using the WS model with 100 nodes and 5,250 links.

Problem 11.1

Let $G_{ER}(n, p)$ be an Erdös–Rényi random network with n nodes and probability p. Use known facts about the spectra of ER random networks to show that if $\bar{k}$ is the average degree of this network then the expected number of triangles tends to $\bar{k}^3/6$ as $n \to \infty$.

For any graph the number of triangles is given by

$$t = \frac{1}{6}\,\mathrm{tr}(A^3) = \frac{1}{6}\sum_{j=1}^{n}\lambda_j^3. \tag{11.6}$$

In an ER graph we know that

$$\lim_{n\to\infty}\lambda_1 = np,\ \ \lim_{n\to\infty}\frac{\lambda_2}{n^{\varepsilon}} = 0,\ \ \lim_{n\to\infty}\frac{\lambda_n}{n^{\varepsilon}} = 0, \tag{11.7}$$

and, since $|\lambda_i| \leq \max\{|\lambda_2|, |\lambda_n|\}$ for $i \geq 1$,

$$\lim_{n \to \infty} t = \frac{1}{6}\lambda_1^3 = \frac{1}{6}(np)^3. \qquad (11.8)$$

Since $\bar{k} = p(n-1)$, as $n \to \infty$,

$$t \to \frac{\bar{k}^3}{6}. \qquad (11.9)$$

Problem 11.2

The data shown in Table 11.1 belong to a network having $n = 1,000$ nodes and $m = 4,000$ links. The network does not have any node with $k \leq 3$. Let $n(k)$ be the number of nodes with degree k. Determine whether this network was generated by the BA model.

The probability that a node chosen at random has a given degree is shown in Table 11.2.

A sketch of the plot of k against $p(k)$ in Figure 11.13 indicates that there is a fast decay of the probability with the degree, which is indicative of fat-tailed degree distributions, like the one produced by the BA model.

If the network was generated with the BA models it has to have a PDF of the form $p(k) \sim k^{-3}$ which means that $\ln p(k) \sim -3\ln k + b$.

Given two degree values k_1 and k_2, the slope of a log–log plot is given by

$$m = \frac{\ln(p(k_2)) - \ln(p(k_1))}{\ln k_2 - \ln k_1}. \qquad (11.10)$$

Using data from the table,

$$m = \frac{\ln(0.003) - \ln(0.196)}{\ln 20 - \ln 5} = -3.0149 \approx -3, \qquad (11.11)$$

indicative of a network generated by the BA model.

....

FURTHER READING

Barabási, A.-L. and Albert, R., *Emergence of scaling in random networks*, *Science* **286**:509–512, 1999.

Bollobás, B., Mathematical results on scale-free random graphs, in Bernholdt, S. and Schuster, H.G. (eds.), *Handbook of Graph and Networks: From the Genome to the Internet*, Wiley-VCH, 1–32, 2003.

Bollobás, B., *Random Graphs*, Cambridge University Press, 2001.

Watts, D.J., Strogatz, S.H., Collective dynamics of 'small-world' networks, *Nature* **393**:440–442, 1998.

Table 11.1 *Degree frequencies in an example network.*

k	n(k)
4	343
5	196
10	23
20	3

Table 11.2 *Probability distribution of degrees in an example network.*

k	p(k)
4	0.343
5	0.196
10	0.023
20	0.003

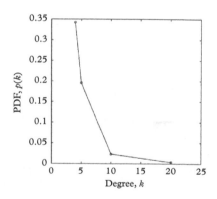

Figure 11.13 *Degree distribution in an illustrative network*

12 Matrix Functions

In this chapter

As we have seen, we can analyse networks by understanding properties of their adjacency matrices. We now introduce some tools for manipulating matrices which will assist in a more detailed analysis, namely functions of matrices. The three most significant in terms of networks will be matrix polynomials, the resolvent, and the matrix exponential and we give a brief introduction to each. We then present elements of a unifying theory for functions of matrices and give examples of some familiar scalar functions in an $n \times n$ dimensional setting.

12.1 Motivation

Amongst one's earliest experiences of mathematics is the application of the basic operations of arithmetic to whole numbers. As we mature mathematically, we see that it is natural to extend the domain of these operations. In particular, we have already exploited the analogues of these operations to develop matrix algebra. Now we look at some familiar functions as applied to matrices. We will exploit these ideas later to better understand networks, for example to develop measures of centrality in Chapter 15 and to measure global properties of networks in Chapters 17 and 18. We start with some familiar ideas that will prove vital in defining a comprehensive generalization.

Note that we will only consider functions of square matrices. Throughout this chapter, assume that $A \in \mathbb{R}^{n \times n}$ unless otherwise stated.[1]

12.2 Matrix powers

We saw matrix powers, A^p when $p \in \mathbb{N}$, when we looked at walks in a network. In this case, A^p is defined simply as the product of p copies of A. To extend the definition, we first let $A^0 = I$ if $A \neq O$.

[1] Almost all results generalize immediately to complex matrices.

Example 12.1

If $D = \text{diag}(\lambda_1, \lambda_2, \ldots, \lambda_n)$ then $D^p = \text{diag}(\lambda_1^p, \lambda_2^p, \ldots, \lambda_n^p)$.

In many applications we need to form A^p for a whole series of values of p. Theoretically, it may be desirable to consider an infinite sequence of powers. The *asymptotic* behaviour of matrix powers is best understood by looking at eigenvalues. Suppose that $A = XJX^{-1}$ where J is the Jordan canonical form of A. Then

$$A^p = \left(XJX^{-1}\right)^p = XJX^{-1}XJX^{-1}XJ \cdots JX^{-1} = XJ^pX^{-1}$$

and the asymptotic behaviour of powers can be understood by considering the powers of J. Assuming that A is simple (which is certainly true for matrices arising from simple networks),

$$A^p = X \begin{bmatrix} \lambda_1^p & & & \\ & \lambda_2^p & & \\ & & \ddots & \\ & & & \lambda_n^p \end{bmatrix} X^{-1}.$$

Clearly, $\lim_{p \to \infty} A^p = O$ if and only if $\rho(A) < 1$ and if $\rho(A) > 1$ then the powers of A grow unboundedly.

If $\rho(A) = 1$ and A is simple then the limiting behaviour is trickier to pin down. While the powers do not grow explosively, $\lim_{p \to \infty} A^p$ exists if and only if the only eigenvalue of size one is one itself.

As implied by the following example, things are more complicated for defective matrices.

Examples 12.2

(i)   Let $J = \begin{bmatrix} \lambda & 1 \\ 0 & \lambda \end{bmatrix}$. Then

$$J^2 = \begin{bmatrix} \lambda^2 & 2\lambda \\ 0 & \lambda^2 \end{bmatrix}, \quad J^3 = \begin{bmatrix} \lambda^3 & 3\lambda^2 \\ 0 & \lambda^3 \end{bmatrix}, \quad J^p = \begin{bmatrix} \lambda^p & p\lambda^{p-1} \\ 0 & \lambda^p \end{bmatrix}.$$

Note that if $|\lambda| < 1$ then $\lim_{p \to \infty} p\lambda^{p-1} = 0$ and the powers converge. If $|\lambda| \geq 1$ they diverge.

continued

Examples 12.2 *continued*

(ii)

$$\text{Let } \mathcal{J} = \begin{bmatrix} 0 & 1 & 0 & 0 \\ 0 & 0 & 1 & 0 \\ 0 & 0 & 0 & 1 \\ 0 & 0 & 0 & 0 \end{bmatrix}. \text{ Then } \mathcal{J}^2 = \begin{bmatrix} 0 & 0 & 1 & 0 \\ 0 & 0 & 0 & 1 \\ 0 & 0 & 0 & 0 \\ 0 & 0 & 0 & 0 \end{bmatrix}, \; \mathcal{J}^3 = \begin{bmatrix} 0 & 0 & 0 & 1 \\ 0 & 0 & 0 & 0 \\ 0 & 0 & 0 & 0 \\ 0 & 0 & 0 & 0 \end{bmatrix}, \; \mathcal{J}^4 = O.$$

Notice that $\rho(\mathcal{J}) = 0$. If all the eigenvalues of a matrix A are zero then it is said to be *nilpotent* and we can show that $A^p = O$ for $p \geq n$.

Once we have defined powers, we can extend the definition of polynomial functions to take matrix arguments. If $p_m(x) = a_0 + a_1 x + \cdots + a_m x^m$ is a degree m polynomial then we define

$$p_m(A) = a_0 I + a_1 A + \cdots + a_m A^m.$$

$p_m(A)$ is defined for all square matrices.

Example 12.3

(i) Let $p(x) = 1 - x^2$ and $A = \begin{bmatrix} 1 & 1 \\ 0 & 2 \end{bmatrix}$. Then $p(A) = I - \begin{bmatrix} 1 & 3 \\ 0 & 4 \end{bmatrix} = \begin{bmatrix} 0 & -3 \\ 0 & -3 \end{bmatrix}$.

(ii) Recall that the *characteristic polynomial* of A is given by $p(z) = \det(A - zI)$. This is a degree n polynomial with the property that $p(z) = 0 \iff z \in \sigma(A)$. The *Cayley–Hamilton theorem* tells us that $p(A) = O$.

If $p(x)$ and $q(x)$ are polynomials then we can define the rational function $r(x) = p(x)/q(x)$ so long as $q(x) \neq 0$. It is natural to write.

$$r(A) = p(A)q(A)^{-1} = q(A)^{-1}p(A),$$

where $q(A)^{-1}$ is the inverse matrix of $q(A)$. This is well defined so long as $q(A)$ is nonsingular.

12.2.1   Resolvent matrix

We are going to need to consider infinite series of matrix powers to get a general understanding of matrix functions. A good place to start is with geometric series.

Examples 12.4

(i) If $z \in \mathbb{C}$ and $z \neq 1$ then

$$1 + z + z^2 + \cdots + z^p = \frac{1 - z^{p+1}}{1 - z}.$$

If $|z| < 1$ then

$$\sum_{i=0}^{\infty} z^i = (1 - z)^{-1}.$$

The function $f(z) = (1 - z)^{-1}$ is an *analytic continuation* of $\sum_{i=0}^{\infty} z^i$ to the punctured disc $\mathbb{C} - \{1\}$.

(ii) $(I + A + A^2 + \cdots + A^p)(I - A) = I - A^{p+1}$. If $(I - A)^{-1}$ exists then we can write

$$I + A + A^2 + \cdots + A^p = (I - A^{p+1})(I - A)^{-1} = (I - A)^{-1}(I - A^{p+1}).$$

$(I - A)$ is invertible so long as 1 is not an eigenvalue of A.

(iii) If $\rho(A) < 1$ then, since $\lim_{p \to \infty} A^p = O$,

$$\sum_{i=0}^{\infty} A^i = (I - A)^{-1}.$$

If $|s| < 1/\rho(A)$ then $\rho(sA) < 1$ and hence

$$s\sum_{i=0}^{\infty} (sA)^i = s(I - sA)^{-1} = (zI - A)^{-1},$$

where $z = 1/s$.

The matrix $(zI - A)^{-1}$ is defined so long as $(zI - A)$ has no zero eigenvalues, which is the case so long as $z \notin \sigma(A)$. Given a matrix A we call

$$f(z) = (zI - A)^{-1}$$

its *resolvent function*. The resolvent function is an analytic continuation of an infinite series of matrix powers to (almost) the whole complex plane. It has a number of applications in network theory but is also useful in extending the notion of matrix functions (even though it is not itself a matrix function).

12.3   The matrix exponential

The matrix exponential arises naturally in the solution of systems of ordinary differential equations: just as the solution to the scalar differential equation

$$\frac{dx}{dt} = kx, \quad x(0) = x_0$$

is $x(t) = x_0 e^{kt}$, so the solution of

$$\frac{d\mathbf{x}}{dt} = A\mathbf{x}, \quad \mathbf{x}(0) = \mathbf{x}_0,$$

where A is an $n \times n$ matrix and $\mathbf{x}(t)$ is a vector of length n, is $\mathbf{x}(t) = e^{tA}\mathbf{x}_0$ where e^{tA} is a *matrix exponential*. This can be defined in a number of ways by generalizing the usual exponential function. Probably the most natural way to write the exponential of a matrix is as the limit of the infinite series

$$e^A = \sum_{k=0}^{\infty} \frac{A^k}{k!}. \tag{12.1}$$

The exponential function of A can be defined as

$$e^{zA} = \sum_{k=0}^{\infty} \frac{(zA)^k}{k!},$$

for all $z \in \mathbb{C}$.

By bounding the size of the terms in this series we can use the comparison test (extended to matrix series) to show that for any square matrix A and any finite $z \in \mathbb{C}$, the series $\sum_{k=0}^{\infty} \frac{(zA)^k}{k!}$ is convergent.

Examples 12.5

(i) $e^{zI} = \sum_{k=0}^{\infty} \frac{(zI)^k}{k!} = \sum_{k=0}^{\infty} \frac{z^k}{k!} I = e^z I.$

(ii) Let $D = \text{diag}(\lambda_1, \lambda_2, \ldots, \lambda_n)$.

$$e^D = \sum_{k=0}^{\infty} \frac{D^k}{k!} = \sum_{k=0}^{\infty} \frac{1}{k!} \begin{bmatrix} \lambda_1^k & & & \\ & \lambda_2^k & & \\ & & \ddots & \\ & & & \lambda_n^k \end{bmatrix} = \begin{bmatrix} e^{\lambda_1} & & & \\ & e^{\lambda_2} & & \\ & & \ddots & \\ & & & e^{\lambda_n} \end{bmatrix}.$$

(iii) If A is symmetric then so is A^k and from (12.1) one can see that e^A is symmetric, too. Again, from (12.1), $e^{A^T} = (e^A)^T$.

Many, but by no means all, of the familiar properties of exponentials extend to the matrix exponential. The following list is not exhaustive. Most can be established by simply inserting the appropriate argument into (12.1).

Theorem 12.1 *Let A and B be square matrices of the same size and let $s, t \in \mathbb{C}$. The following properties hold for matrix exponentials.*

1. $e^O = I$.
2. $\dfrac{d}{dt}\left(e^{tA}\right) = Ae^{tA}$.
3. $e^{(s+t)A} = e^{sA}e^{tA}$.
4. $e^{A+B} = e^A e^B$ if and only if $AB = BA$.
5. e^{tA} is nonsingular and $\left(e^{tA}\right)^{-1} = e^{-tA}$.
6. If B is nonsingular then $Be^{tA}B^{-1} = e^{tBAB^{-1}}$.
7. $e^A = \lim\limits_{k \to \infty} (I + A/k)^k$.
8. $e^{tA} = \left(e^A\right)^t$.

Problem 12.1

Show that if $AB = BA$ then $e^{A+B} = e^A e^B$.

By a simple induction, $AB^k = B^k A$ for any k and hence from (12.1) $Ae^{tB} = e^{tB}A$. Similarly, $Xe^{tY} = e^{tY}X$ for any combination of X and Y chosen from A, B, and $A + B$.

Now let $F(t) = e^{t(A+B)}e^{-Bt}e^{-At}$. By the product rule,

$$F'(t) = (A + B)e^{(A+B)t}e^{-Bt}e^{-At} + e^{(A+B)t}(-B)e^{-Bt}e^{-At} + e^{(A+B)t}e^{-Bt}(-A)e^{-At},$$

and by commutativity, the right-hand side of this expression is zero. Thus $F(t)$ is a constant matrix and since $F(0) = e^O e^O e^O = I$ we know that

$$e^{t(A+B)} = e^{tA}e^{tB}$$

for all t (in particular when $t = 1$).

Examples 12.6

(i) Let $\mathcal{J} = \begin{bmatrix} 0 & 1 & 0 \\ 0 & 0 & 1 \\ 0 & 0 & 0 \end{bmatrix}$. Then, $\mathcal{J}^2 = \begin{bmatrix} 0 & 0 & 1 \\ 0 & 0 & 0 \\ 0 & 0 & 0 \end{bmatrix}$ and $\mathcal{J}^3 = O$ so $e^{\mathcal{J}} = \begin{bmatrix} 1 & 1 & \frac{1}{2} \\ 0 & 1 & 1 \\ 0 & 0 & 1 \end{bmatrix}$.

(ii) If $A = \begin{bmatrix} \lambda & 1 & 0 \\ 0 & \lambda & 1 \\ 0 & 0 & \lambda \end{bmatrix}$ then $A = \lambda I + \mathcal{J}$ (where $\mathcal{J}$ is taken from (i)). So

$$e^A = e^{\lambda I + \mathcal{J}} = e^{\lambda I}e^{\mathcal{J}} = e^{\lambda}\begin{bmatrix} 1 & 1 & \frac{1}{2} \\ 0 & 1 & 1 \\ 0 & 0 & 1 \end{bmatrix}.$$

Examples 12.6 *continued*

(iii) Let $A = \begin{bmatrix} 0 & 1 \\ 0 & 1 \end{bmatrix}$. It is easy to show that $A^k = A$ for all k. So

$$e^A = I + \sum_{k=1}^{\infty} \frac{A}{k!} = \begin{bmatrix} 1 & \sum_{k=1}^{\infty} 1/k! \\ 0 & \sum_{k=0}^{\infty} 1/k! \end{bmatrix} = \begin{bmatrix} 1 & e-1 \\ 0 & e \end{bmatrix}.$$

Write $A = D + F$ where $D = \begin{bmatrix} 0 & 0 \\ 0 & 1 \end{bmatrix}, F = \begin{bmatrix} 0 & 1 \\ 0 & 0 \end{bmatrix}$.

Since D is diagonal and $F^2 = O$, $e^D = \begin{bmatrix} 1 & 0 \\ 0 & e \end{bmatrix}$ and $e^F = \begin{bmatrix} 1 & 1 \\ 0 & 1 \end{bmatrix}$.

So $e^D e^F = \begin{bmatrix} 1 & 1 \\ 0 & e \end{bmatrix}$ and $e^F e^D = \begin{bmatrix} 1 & e \\ 0 & e \end{bmatrix}$. Neither of these equal e^A since $DF \neq FD$.

Suppose that we have computed the Jordan decomposition of a matrix: $A = XJX^{-1}$. Using Theorem 12.1 we have

$$e^{tA} = e^{tXJX^{-1}} = Xe^{tJ}X^{-1},$$

and if we know e^{tJ} we can use it to compute e^{tA}. In particular, if A is simple we can take advantage of the fact that computation of the exponential of a diagonal matrix is particularly straightforward.

Examples 12.7

(i) $A = \begin{bmatrix} 5 & 3 \\ -6 & -4 \end{bmatrix} = \begin{bmatrix} 1 & 1 \\ -1 & -2 \end{bmatrix} \begin{bmatrix} 2 & 0 \\ 0 & -1 \end{bmatrix} \begin{bmatrix} 1 & 1 \\ -1 & -2 \end{bmatrix}^{-1}$ hence

$$e^A = \begin{bmatrix} 1 & 1 \\ -1 & -2 \end{bmatrix} \begin{bmatrix} e^2 & 0 \\ 0 & e^{-1} \end{bmatrix} \begin{bmatrix} 2 & 1 \\ -1 & -1 \end{bmatrix} = \begin{bmatrix} 2e^2 - e^{-1} & e^2 - e-1 \\ 2e^{-1} - 2e^2 & 2e^{-1} - e^2 \end{bmatrix}$$

and $e^{tA} = \begin{bmatrix} 2e^{2t} - e^{-t} & e^{2t} - e-t \\ 2e^{-t} - 2e^{2t} & 2e^{-t} - e^{2t} \end{bmatrix}$.

(ii) $B = \begin{bmatrix} -5 & 2 \\ -15 & 6 \end{bmatrix} = \begin{bmatrix} 1 & 2 \\ 3 & 5 \end{bmatrix} \begin{bmatrix} 1 & 0 \\ 0 & 0 \end{bmatrix} \begin{bmatrix} 1 & 2 \\ 3 & 5 \end{bmatrix}^{-1}$ hence

$$e^B = \begin{bmatrix} 1 & 2 \\ 3 & 5 \end{bmatrix} \begin{bmatrix} e & 0 \\ 0 & 1 \end{bmatrix} \begin{bmatrix} -5 & 2 \\ 3 & -1 \end{bmatrix} = \begin{bmatrix} 6 - 5e & 2e - 2 \\ 15 - 15e & 6e - 5 \end{bmatrix}.$$

(iii) $e^A e^B = \begin{bmatrix} 2e^2 - e^{-1} & e^2 - e - 1 \\ 2e^{-1} - 2e^2 & 2e^{-1} - e^2 \end{bmatrix} \begin{bmatrix} 6 - 5e & 2e - 2 \\ 15 - 15e & 6e - 5 \end{bmatrix} = \begin{bmatrix} -290.36 & 128.93 \\ 278.08 & -123.50 \end{bmatrix}$. We can show that

$e^{A+B} = \begin{bmatrix} -1.759 & -0.931 \\ 3.910 & -2.131 \end{bmatrix}$.

As is the case for matrix powers, the asymptotic behaviour of e^{tA} is linked directly to the spectrum of A. If A is simple then

$$e^{tA} = X \begin{bmatrix} e^{\lambda_1 t} & & & \\ & e^{\lambda_2 t} & & \\ & & \ddots & \\ & & & e^{\lambda_n t} \end{bmatrix} X^{-1}, \tag{12.2}$$

where the λ_i are the eigenvalues of A (and X contains the eigenvectors). We can infer that if A is simple then $\lim_{t \to \infty} e^{tA} = O$ if and only if all the eigenvalues of A lie to the left of the imaginary axis in the complex plane; that the limit diverges if any of the eigenvalues lie strictly to the right, and that things are more complicated when eigenvalues lie on (but not to the right of) the imaginary axis.

The matrix exponential has a number of roles in network theory. As an example, consider the following.

Problem 12.2
Show that a network with adjacency matrix A is connected if and only if $e^A > 0$.

If there is a walk of length p between nodes i and j then there is a nonzero in the (i,j)th element of A^p. If a network is connected then a nonzero must eventually appear in the sequence

$$a_{ij}, (A^2)_{ij}, (A^3)_{ij}, \ldots$$

for every i and j. Recall that $e^A = \sum_{k=0}^{\infty} \frac{A^k}{k!}$ and a nonzero must therefore appear in every element of the series at some point. Since A is nonnegative, every nonzero entry gives a positive contribution to the final value of e^A.

Conversely, if $e^A > 0$ then we can trace every nonzero entry of the exponential to a nonzero in the series and hence to a path in the network.

12.4 Extending matrix powers

Just as for numbers, the power of a matrix, A^p, can be extended to have meaning for any $p \in \mathbb{R}$ (and even for complex powers, too). For negative p we simply use the matrix inverse: for $p \in \mathbb{N}, A^{-p} = (A^{-1})^p$. This is well defined so long as A is non-singular. For rational powers we need to extend the idea of pth roots to matrices.

12.4.1 The matrix square root

If $z \in \mathbb{C}$ is nonzero then there are two distinct numbers, ω_0 and ω_1 such that $\omega_i^2 = z$. These two numbers are the square roots of z and $\omega_0 = -\omega_1$. A naïve approach to extending the idea of the square root to a matrix A is to look for solutions to the equation $X^2 = A$. A brief consideration of this problem shows some of the complications of extending the definition of functions to accept matrix arguments.

Examples 12.8

(i) Let $A = \begin{bmatrix} 0 & 1 \\ 0 & 0 \end{bmatrix}$ and suppose $X^2 = A$. Since A is nilpotent then so is X. But then $X^2 = O$, a contradiction, so A has no square roots.

   Notice that A is one of the (infinite) solutions to the equation $X^2 = O$.

(ii) Suppose $A = SDS^{-1}$ where $D = \operatorname{diag}(\lambda_1, \lambda_2, \ldots, \lambda_n)$ and let $E = \operatorname{diag}(\mu_1, \mu_2, \ldots, \mu_n)$ where μ_i is one of the square roots of λ_i. Then

$$\left(SES^{-1} \right)^2 = SE^2S^{-1} = A.$$

   If $\lambda_i \neq 0$ then there are two choices for μ_i and by going through the various permutations we get (up to) 2^n different solutions to $X^2 = A$.

(iii) If $X^2 = A$ then $(-X)^2 = A$, too.

(iv) The only two solutions to $X^2 = \begin{bmatrix} 1 & 1 \\ 0 & 1 \end{bmatrix}$ are $X = \pm \begin{bmatrix} 1 & \frac{1}{2} \\ 0 & 1 \end{bmatrix}$.

From the examples we see that the matrix equation $X^2 = A$ exhibits an unexpectedly complicated behaviour compared to the scalar analogue, particularly for defective matrices. To define a matrix square root function we need to give a unique value to $\sqrt{A}$. Recall that the principal square root of $z \in \mathbb{C}$, written $\sqrt{z}$, is the solution of $\omega^2 = z$ with smallest principal argument. If z is real and positive then so is $\sqrt{z}$.

If A is simple and has the factorization $A = XDX^{-1}$ where $D = \operatorname{diag}(\lambda_1, \ldots, \lambda_n)$ then the *principal square root* of A is the unique matrix

$$\sqrt{A} = X\operatorname{diag}(\sqrt{\lambda_1}, \ldots, \sqrt{\lambda_n})X^{-1}. \tag{12.3}$$

If A is defective then the square root function, $\sqrt{A}$, is not defined. If all the eigenvalues of a simple matrix are real and positive then the same is true of its principal square root. For most applications, the principal square root is the right one to take. Having defined the principal square root, there is an obvious extension to pth roots by replacing $\sqrt{\lambda_i}$ in (12.3) with $\lambda_i^{1/p}$.

Combining pth powers, qth roots, and the inverse, one can define a principal rth power of a matrix for any $r \in \mathbb{Q}$. For irrational values of r we need to make use of the matrix exponential in a similar way to the extension of irrational powers of scalars.

Example 12.9

If $D = \operatorname{diag}(\lambda_1, \lambda_2, \ldots, \lambda_n)$ then $e^D = \operatorname{diag}(e^{\lambda_1}, e^{\lambda_2}, \ldots, e^{\lambda_n})$. If $p \in \mathbb{Q}$ then

$$\left(e^D\right)^p = \operatorname{diag}(e^{\lambda_1}, e^{\lambda_2}, \ldots, e^{\lambda_n})^p = \operatorname{diag}(e^{p\lambda_1}, e^{p\lambda_2}, \ldots, e^{p\lambda_n}) = e^{pD},$$

where we take the principal value of $(e^{\lambda_i})^p$ in each case.

If A is simple we can extend the last example using the Jordan decomposition to show that $e^{pA} = \left(e^A\right)^p$ for rational values of p. We can extend the definition of matrix powers to irrational indices by saying $A^t = e^{tX}$ where $e^X = A$.[2]

12.5 General matrix functions

Equations (12.1) and (12.2) suggest two ways of extending the definition of any analytic function to take matrix arguments. Suppose $f : \mathbb{C} \to \mathbb{C}$ is analytic on some domain with Maclaurin series

$$f(z) = \sum_{k=0}^{\infty} a_k z^k$$

then define $f : \mathbb{R}^{n \times n} \to \mathbb{R}^{n \times n}$ by

$$f(A) = \sum_{k=0}^{\infty} a_k A^k. \tag{12.4}$$

Alternatively, if A is simple with Jordan decomposition XDX^{-1} where $D = \operatorname{diag}(\lambda_1, \ldots, \lambda_n)$ then define

$$f(A) = X \operatorname{diag}(f(\lambda_1), \ldots, f(\lambda_n)) X^{-1}. \tag{12.5}$$

There are other ways of extending the definition of scalar functions to matrices, too. We will give just one more. If you have studied complex analysis you may recall that Cauchy's integral formula tells us that if f is analytic inside some region R and $a \in R$ then

---

[2] But how do we find X?

$$f(a) = \frac{1}{2\pi i} \oint_C \frac{f(z)}{z-a} dz$$

where C is a circle in R containing a. For matrix functions this becomes

$$f(A) = \frac{1}{2\pi i} \oint_C f(z)(zI - A)^{-1} dz \qquad (12.6)$$

for an appropriate contour C: notice the role of the resolvent function. We interpret the integral as being over each of the n^2 elements of the function $f(z)(zI - A)^{-1}$. But what contour should we use? We need to enclose all the poles of the resolvent, so C needs to contain all the eigenvalues of A. The contour integral formula can be useful for reducing the length of proofs of certain theoretical results and allows matrix functions to be generalized even further to operators. Recently, computational techniques for computing matrix functions that exploit the integral formula have been developed. These make use of quadrature rules to perform the integrations.

If A is a simple matrix then

$$X\text{diag}(f(\lambda_1),\ldots,f(\lambda_n))X^{-1} = X\text{diag}\left(\sum_{k=0}^{\infty} a_k\lambda_1^k, \ldots, \sum_{k=0}^{\infty} a_k\lambda_n^k\right) X^{-1}$$

$$= X\left[\sum_{k=0}^{\infty} a_k\text{diag}\left(\lambda_1^k,\ldots,\lambda_n^k\right)\right] X^{-1}$$

$$= \sum_{k=0}^{\infty} a_k XD^k X^{-1} = \sum_{k=0}^{\infty} a_k A^k,$$

so (12.4) and (12.5) are equivalent.

Problem 12.3
Show that if $f(z)$ is analytic in a region R containing the spectrum of A then (12.5) and (12.6) are equivalent.

If $z \notin \sigma(A)$ then $(zI - A)^{-1}$ is well defined and $(zI - A)^{-1} = X(zI - D)^{-1}X^{-1}$. Note that $(zI - D)^{-1}$ is diagonal with entries of the form $1/(z - \lambda_i)$. So, since $\sigma(A) \bigcap C = \varnothing$,

$$\frac{1}{2\pi i} \oint_C f(z)(zI - A)^{-1} dz = \frac{1}{2\pi i} X\left(\oint_C f(z)(zI - D)^{-1} dz\right) X^{-1}$$

$$= \frac{1}{2\pi i} X\left(\oint_C f(z)\begin{bmatrix} \frac{1}{z-\lambda_1} & & \\ & \ddots & \\ & & \frac{1}{z-\lambda_n} \end{bmatrix} dz\right) X^{-1}$$

$$= X \begin{bmatrix} \dfrac{1}{2\pi i} \oint_C \dfrac{f(z)}{z-\lambda_1}\,dz & & \\ & \ddots & \\ & & \dfrac{1}{2\pi i} \oint_C \dfrac{f(z)}{z-\lambda_n}\,dz \end{bmatrix} X^{-1}$$

$$= X \begin{bmatrix} f(\lambda_1) & & \\ & \ddots & \\ & & f(\lambda_n) \end{bmatrix} X^{-1}.$$

It can be shown using some basic tools of analysis that (12.4) and (12.6) are also equivalent for defective matrices.[3] If $\mathcal{J}$ is a $(p+1) \times (p+1)$ defective Jordan block of the eigenvalue λ then one can show that if we define

$$f(\mathcal{J}) = \begin{bmatrix} f(\lambda) & f'(\lambda) & \cdots & \dfrac{f^{(p)}(\lambda)}{p!} \\ & f(\lambda) & & \vdots \\ & & \ddots & f'(\lambda) \\ & & & f(\lambda) \end{bmatrix} \tag{12.7}$$

then the consequent extension of (12.5) is equivalent to the other definitions, too.

Examples 12.10

(i) If $f(z) = \sqrt{z}$ then $f'(z)$ is undefined for $z = 0$.

   If $\mathcal{J} = \begin{bmatrix} 0 & 1 \\ 0 & 0 \end{bmatrix}$ then (12.7) is undefined, there is no Maclaurin series for $f(z)$ and the conditions for the Cauchy integral formula are not met because any contour enclosing $\sigma(\mathcal{J})$ includes zero. Recall that $\mathcal{J}$ has no square root.

(ii) If $A = O$ then we still cannot use (12.4) or (12.6) to define $\sqrt{A}$, but (12.5) works fine.

If $f(z)$ is analytic at $z \in \sigma(A)$ then the value of $f(A)$ given by our equivalent definitions is called the *primary value* of $f(A)$. Applications for values other than the primary are limited and we will not consider them further.

Using (12.4) we can establish a number of generic properties that any matrix function satisfies. If f is analytic on the spectrum of A then, for example, $f(A^T) = f(A)^T$ and $f(XAX^{-1}) = Xf(A)X^{-1}$.

[3] One uses the fact that simple matrices form a dense set amongst all square matrices.

Problem 12.4

Show that if $AB = BA$ and f is analytic on the spectrum of A and B then $f(A)A = Af(A), f(A)B = Bf(A)$, and $f(A)f(B) = f(B)f(A)$.

$$f(A)A = \left(\sum_{m=0}^{\infty} a_m A^m\right) A = \sum_{m=0}^{\infty} a_m A^{m+1} = A\left(\sum_{m=0}^{\infty} a_m A^m\right) = Af(A).$$

Also, since $A^m B = A^{m-1}AB = A^{m-1}BA = A^{m-2}ABA = \cdots = BA^m$,

$$f(A)B = \left(\sum_{m=0}^{\infty} a_m A^m\right) B = \sum_{m=0}^{\infty} a_m (A^m B) = \sum_{m=0}^{\infty} a_m (BA^m) = B\left(\sum_{m=0}^{\infty} a_m A^m\right) = Bf(A).$$

Finally, write $C = f(B)$. We have just shown that $AC = CA$ hence

$$f(A)f(B) = f(A)C = Cf(A) = f(B)f(A).$$

We finish the chapter by defining matrix versions of some familiar functions.

12.5.1 $\log A$

X is a logarithm of A if $e^X = A$. Suppose $A = XDX^{-1}$ where $D = \operatorname{diag}(\lambda_1, \ldots, \lambda_n)$. Since A is nonsingular, all the λ_j are nonzero and if

$$\xi_j = \log|\lambda_j| + i\operatorname{Arg}(\lambda_j) + 2\pi k_j i,$$

for any $k_j \in \mathbb{Z}$ then $e^{\xi_j} = \lambda_j$. For any choice of k_j let $L = \operatorname{diag}(\xi_1, \ldots, \xi_n)$. Using (12.5) we can show that $e^{XLX^{-1}} = A$. Hence every nonsingular simple matrix has an infinite number of logarithms. We can show that if A is singular then there are no solutions to $e^X = A$.

Of the infinite number of possibilities for nonsingular matrices, it is very unusual to use anything other than the primary logarithm defined by (12.5). To ensure continuity, one often restricts the definition of the logarithm to matrices without eigenvalues on the negative real axis. The matrix logarithm has a number of applications when coupled with the matrix exponential. As $\log z$ is not defined at zero, we cannot use (12.4) but we can derive power series representations by using formulae for the scalar logarithm.

Examples 12.11

(i) If $|z| < 1$ then

$$\log(1 + z) = z + \frac{z^2}{2} + \frac{z^3}{3} + \cdots.$$

If $\rho(A) < 1$ and A is nonsingular then one can show that

$$\log(I + A) = A + \frac{1}{2}A^2 + \frac{1}{3}A^3 + \cdots .$$

(ii) Suppose $X = \log A$, $Y = \log B$, and $AB = BA$. By Theorem 12.1, $XY = YX$ and so by Theorem 12.1 $e^X e^Y = e^{X+Y}$.
  Therefore if $AB = BA$, $AB = e^{\log A + \log B}$. Taking logs of each side gives

$$\log(AB) = \log A + \log B.$$

Strictly speaking, we have not shown that each of the logs in this expression is the primary value, but there is a large class of matrices for which this is true.

(iii) Let $X = \log A$. By Theorem 12.1 $A^p = (e^X)^p = e^{pX}$.
  Taking logs of each side gives

$$\log A^p = p \log A.$$

In particular, $\log A^{-1} = -\log A$.

12.5.2 $\cos A$ and $\sin A$

We can define $\cos A$ and $\sin A$ for any matrix using the Maclaurin series for the scalar functions. Alternatively, we can use Euler's formula

$$e^{i\theta} = \cos \theta + i \sin \theta$$

and write

$$\cos A = \frac{1}{2}(e^{iA} + e^{-iA}), \quad \sin A = \frac{1}{2i}(e^{iA} - e^{-iA}).$$

We can use these identities to show that many trigonometric identities still hold when we use matrix arguments. For example, the addition formulae for cosine and sine are true for matrices *so long as* the matrices involved commute.

Example 12.12

$$\cos^2 A + \sin^2 A = (\cos A + i \sin A)(\cos A - i \sin A) = e^{iA}e^{-iA} = I.$$

As with there scalar equivalents, $\cos A$ and $\sin A$ arise naturally in the solution of (systems of) second order ODEs. They are far removed from their original role in trigonometry!

We can define hyperbolic functions of matrices, too.

$$\cosh A = \frac{1}{2}(e^A + e^{-A}), \quad \sinh A = \frac{1}{2}(e^A - e^{-A}).$$

Examples 12.13

(i) Using the power series definition of e^A we can write

$$\sinh A = \frac{e^A - e^{-A}}{2} = \sum_{k=0}^{\infty} \frac{A^{2k+1}}{(2k+1)!}.$$

Recall that in a bipartite network there are no odd cycles. This manifests itself as a zero diagonal in $\sinh A$ where A is the adjacency matrix.

(ii) Similarly,

$$\cosh A = \sum_{k=0}^{\infty} \frac{A^{2k}}{(2k)!}.$$

A comparison between cosh and sinh of adjacency matrices gives insight into the weighting of odd and even walks.

FURTHER READING

Benzi, M. and Boito, P. Quadrature rule-based bounds for functions of adjacency matrices, Lin. Alg. Appl. **433**:637–652, 2010.

Estrada, E. and Higham, D.J., Network properties revealed through matrix functions, SIAM Rev., **52**:696–714, 2010.

Higham, N.J., *Functions of Matrices*, SIAM, 2008.

Moler, C. and Van Loan, C., Nineteen Dubious Ways to Compute the Exponential of a Matrix, Twenty-Five Years Later, SIAM Rev., **45**:3–49, 2003.

Fragment-based Measures

13

In this chapter

We start with the definition of a fragment, or subgraph, in a network. We then introduce the concept of network motif and analyse how to quantify its significance. We illustrate the concept by studying motifs in some real-world networks. We then outline mathematical methods to quantify the number of small subgraphs in networks analytically. We develop some general techniques that can be adapted to search for other fragments.

13.1 Motivation

In many real-life situations, we are able to identify small structural pieces of a system which are responsible for certain functional properties of the whole system. Biologists, chemists, and engineers usually isolate these small fragments of the system to understand how they work and gain understanding of their roles in the whole system. These kinds of structural fragments exist in complex networks. In Chapter 11 we saw that triangles can indicate transitive relations in social networks. They also play a role in interactions between other entities in complex systems. In this chapter we develop techniques to quantify some of the simplest but most important fragments or subgraphs in networks. We also show how to determine whether the presence of these fragments in a real-world network is just a manifestation of a random underlying process, or that they signify something more significant.

In network theory fragments are synonymous with subgraphs. Typical subgraphs are illustrated in Figure 13.1. In general, a subgraph can be formed by one (connected subgraph) or more (disconnected subgraphs) connected components, and they may be cyclic or acyclic.

13.2 Counting subgraphs in networks

In order to count subgraphs in a network we need to use a combination of algebraic and combinatorial techniques. First, we are going to develop some basic techniques which can be combined to count many different types of subgraph.

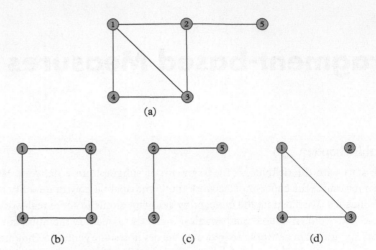

Figure 13.1 *Three subgraphs (bottom line) in an undirected network (top)*

13.2.1 Counting stars

We know from previous chapters that the number of edges in a network can be obtained from the degrees of the nodes. An edge is a star subgraph of the type $S_{1,1}$. Thus

$$|S_{1,1}| = m = \frac{1}{2} \sum_{i=1}^{n} k_i. \tag{13.1}$$

The next star subgraph is $S_{1,2}$. Copies of $S_{1,2}$ in a network can be enumerated by noting that they are formed from any two edges incident to a common node. That is, $|S_{1,2}|$ is equal to the number of times that the nodes attached to a particular node can be combined in pairs. This is simply

$$|S_{1,2}| = \sum_{i=1}^{n} \binom{k_i}{2} = \frac{1}{2} \sum_{i=1}^{n} k_i(k_i - 1). \tag{13.2}$$

Since $S_{1,2}$ is the same as P_2, we have generated the formula we used in calculating transitivity in Chapter 10.

Similarly, the number of $S_{1,3}$ star subgraphs equals the number of times that the nodes attached to a given node can be combined in triples, namely,

$$|S_{1,3}| = \sum_{i=1}^{n} \binom{k_i}{3} = \frac{1}{6} \sum_{i=1}^{n} k_i(k_i - 1)(k_i - 2). \tag{13.3}$$

In general, the number of star subgraphs of the type $S_{1,s}$ is given by

$$|S_{1,s}| = \sum_{i=1}^{n} \binom{k_i}{s}. \tag{13.4}$$

13.2.2 Using closed walks

The idea of using closed walks (CWs) to count subgraphs is very intuitive and simple. Every time that we complete a CW in a network we have visited a sequence of nodes and edges which together form a subgraph. For instance, a CW that goes from a node to any of its neighbours and back again describes an edge, while every CW of length three necessarily visits all the nodes of a triangle. Note that CWs of length $l = 2d, d = 1, 2, \ldots$, that go back and forth between adjacent nodes also describe edges. Similarly, a CW of length $l = 2d + 1, d = 1, 2, \ldots$, visiting only three nodes of the network describes a triangle. Keeping this in mind, we can design a technique to express the number of CWs as a sum of fragment contributions. We start by designating by μ_l the number of CWs of length l in a network. Then $\mu_0 = n$ and, in a simple network, $\mu_1 = 0$. In general, we know that

$$\mu_k = \text{tr}(A^k) = \sum_{i=1}^{n} \lambda_i^k$$

but we can rewrite the right-hand side of this expression in terms of particular subgraphs. Before continuing, visualize what happens with a CW of length two. Each such walk represents an edge. But in an undirected network there are two closed walks along each edge (i,j), namely $i \to j \to i$ and $j \to i \to j$. Thus,

$$\mu_2 = 2|S_{1,1}| = 2|P_1|.$$

Similarly, there are six CWs of length three around every triangle $\triangle_{i,j,k}$ since we can start from any one of its three nodes and move either clockwise or anticlockwise: $i \to j \to k \to i$; $i \to k \to j \to i$; $j \to k \to i \to j$; $j \to i \to k \to j$; $k \to i \to j \to k$; $k \to j \to i \to k$. So,

$$\mu_3 = 6|C_3|.$$

Things begin to get messy for longer CWs as there are several subgraphs associated with such walks. For example, a CW of length four can be generated by moving along the same edge four times. This can be done in two ways

$$i \to j \to i \to j \to i \text{ and } j \to i \to j \to i \to j.$$

We could also walk along two edges and then return to the origin in two ways,

$$i \to j \to k \to j \to i \text{ and } k \to j \to i \to j \to k.$$

There are two ways of visiting two nearest neighbours before returning to the origin,

$$j \to i \to j \to k \to j \text{ and } j \to k \to j \to i \to j.$$

And finally, there are eight ways of completing a cycle of length four in a square i, j, k, l since we can start from any node and go clockwise or anticlockwise. For example, starting from node i gives

$$i \to j \to k \to l \to i \text{ and } i \to l \to k \to j \to i.$$

Consequently,

$$\mu_4 = 2|P_1| + 4|P_2| + 8|C_4|.$$

Problem 13.1

Let G be a regular network with $n = 2r$ nodes of degree k and spectrum

$$\sigma(G) = \{[k]^1, [1]^{r-1}, [-1]^{r-1}, [-k]^1\}. \tag{13.5}$$

Find expressions for the number of triangles and squares in G.

The number of triangles in a network is given by

$$t = \frac{1}{6}\text{tr}(A^3) = \frac{1}{6}\sum_{i=1}^{n}\lambda_i^3 = \frac{1}{6}\left(k^3 + (-k)^3 + (r-1)\left(1^3 + (-1)^3\right)\right) = 0.$$

The number of squares is given by $|C_4| = \mu_4/8 - |P_1|/4 - |P_2|/2$. Since each node has degree k,

$$|P_1| = \frac{kn}{2} = kr$$

and

$$|P_2| = \sum_{i=1}^{n}\binom{k}{2} = \frac{nk(k-1)}{2} = nk(k-1). \tag{13.6}$$

Given that

$$\mu_4 = \text{tr}(A^4) = k^4 + (-k)^4 + (r-1) + (r-1)(-1)^4 = 2k^4 + 2(r-1), \tag{13.7}$$

we conclude that

$$|C_4| = \frac{k^4 + (r-1)}{4} - \frac{rk(k-1)}{2} - \frac{rk}{4} = \frac{k^4 + r - 1 - 2rk(k-1) - rk}{4}$$

$$= \frac{k^4 + r - rk(2k-1) - 1}{4}. \tag{13.8}$$

Example 13.1

A network of the type described in problem 13.2 is the cube Q_3 which has the spectrum

$$\sigma(G) = \{[3]^1, [1]^3, [-1]^3, [-3]^1\}.$$

Applying the formula obtained for the number of squares gives

$$|C_4| = \frac{3^4 + 4 - 60 - 1}{4} = 6,$$

which is the number of faces on a cube.

13.2.3 Combined techniques

We start by considering a practical example. In this case, we will be interested in computing the number of fragments of the type illustrated in Figure 13.2. This fragment is known as a tadpole $T_{3,1}$ subgraph.

The fragment $T_{3,1}$ is characterized by having a node which is simultaneously part of a triangle and of a path of length one. We can combine the idea of calculating the number of triangles in which the node is involved with the number of nodes attached to it. Let t_i be the number of triangles attached to the node i and let $k_i > 2$ be the degree of this node. The number of nodes not in the triangle which are attached to i is just the remaining degree of the node, i.e. $k_i - 2$. Thus, the number of tadpole subgraphs in which the node i is involved is $k_i - 2$ times its number of triangles. Consequently,

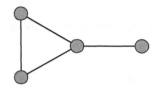

Figure 13.2 *The tadpole graph $T_{3,1}$*

$$|T_{3,1}| = \sum_{k_i > 2} t_i(k_i - 2). \tag{13.9}$$

Example 13.2

Find the number of fragments $T_{3,1}$ in Figure 13.3.

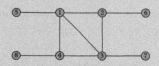

Figure 13.3 *A network with many tadpole fragments*

continued

Example 13.2 *continued*

Using (13.9) and concentrating only on those nodes with degree larger than two we have,

$$|T_{3,1}| = 2 \times (4-2) + 1 \times (3-2) + 2 \times (4-2) + 1 \times (3-2) = 10.$$

Can you see all of them?

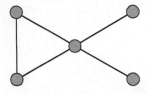

Figure 13.4 *Illustration of the cricket graph*

Let us add another edge and consider the fragment illustrated in Figure 13.4.

This subgraph is known as the cricket graph, which we designate by Cr. Here again, we can use a technique that combines the calculation of the two subgraphs forming this fragment. That is, this fragment is characterized by a node i that is simultaneously part of a triangle and a star $S_{1,2}$. Using t_i as before, we consider nodes for which $k_i > 3$.

If $t_i > 0$ then node i has $k_i - 2$ additional nodes which are attached to it. These $k_i - 2$ nodes can be combined in $\binom{k_i - 2}{2}$ pairs to form all the $S_{1,2}$ subgraphs in which node i is involved.

The number of crickets involving node i is then

$$|Cr_i| = t_i \binom{k_i - 2}{2} \tag{13.10}$$

and hence

$$|Cr| = \sum_{k_i \geq 4} t_i \binom{k_i - 2}{2} = \frac{1}{2} \sum_{k_i \geq 4} t_i (k_i - 2) \cdot (k_i - 3). \tag{13.11}$$

Example 13.3

Find the number of Cr fragments in the network illustrated in Figure 13.3.

Only the nodes labelled as one and three have degree larger than or equal to four. We have,

$$|Cr| = \frac{1}{2}(2(4-2)(4-3) + 2(4-2)(4-3)) = 4. \tag{13.12}$$

The four crickets are illustrated in Figure 13.5:

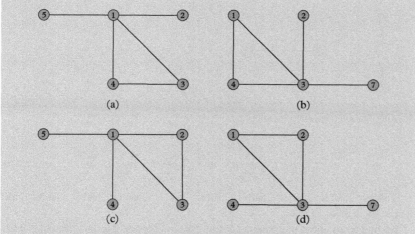

Figure 13.5 *Illustration of the four cricket subgraphs within the network in figure 13.3*

13.2.4 Other techniques

The diamond graph (D) is characterized by the existence of two connected nodes (1 and 3) which are also connected by two paths of length two (1-2-3 and 1-4-3). It is illustrated in Figure 13.6. To calculate the number of diamonds in a network we note that the number of walks of length two between two *connected* nodes is given by $(A^2)_{ij}A_{ij}$ and hence that the number of pairs of paths of length two among two connected nodes i, j is given by

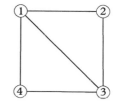

Figure 13.6 *The diamond graph*

$$\binom{(A^2)_{ij}A_{ij}}{2}.$$

Consequently, the number of diamond subgraphs in a network is given by

$$|D| = \frac{1}{2} \sum_{i,j} \binom{(A^2)_{ij}A_{ij}}{2} = \frac{1}{4} \sum_{i,j} \left((A^2)_{ij}A_{ij}\right)\left((A^2)_{ij}A_{ij} - 1\right). \qquad (13.13)$$

Problem 13.2
Find an expression for $|C_5|$, the number of pentagons in a network.

A CW of length $l = 2d+1$ necessarily visits only nodes in subgraphs containing at least one odd cycle. So a CW of length five can visit only the nodes of a triangle, C_3; a tadpole, $T_{3,1}$; or a pentagon, C_5. Hence

$$\mu_5 = a|C_3| + b|T_{3,1}| + c|C_5| \qquad (13.14)$$

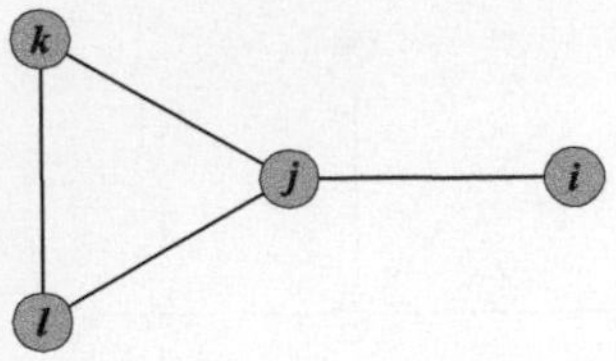

$$i \to j \to k \to l \to j \to i \qquad j \to i \to j \to l \to k \to j$$
$$i \to j \to l \to k \to j \to i \qquad k \to l \to j \to i \to j \to k$$
$$j \to k \to l \to j \to i \to j \qquad k \to j \to i \to j \to l \to k$$
$$j \to l \to k \to j \to i \to j \qquad l \to j \to i \to j \to k \to l$$
$$j \to i \to j \to k \to l \to j \qquad l \to k \to j \to i \to j \to l$$

Figure 13.7 *CWs of length five in $T_{3,1}$*

and

$$|C_5| = \frac{1}{c}(\mu_5 - a|C_3| - b|T_{3,1}|). \tag{13.15}$$

We have seen how to calculate $|C_3|$ and $|T_{3,1}|$ hence our task is to determine the coefficients a, b, and c.

To find a we must enumerate all the CWs of length five in a triangle. This can be done by calculating $\mathrm{tr}(A_C^5)$ where A_C is the adjacency matrix of C_3. From Chapter 6 we know that the eigenvalues of C_3 are 2, –1, and –1 hence

$$a = \sum_i \lambda_j^5 = 2^5 + 2(-1)^5 = 30. \tag{13.16}$$

To find b we can enumerate all the CWs of length five involving all the nodes of $T_{3,1}$. This is done in Figure 13.7 and we see that $b = 10$.

We could also proceed in a similar way as for the triangle, but in the tadpole $T_{3,1}$ not every CW of length five visits all the nodes of the fragment. That is, there are CWs of length five which only visit the nodes of the triangle in $T_{3,1}$. Thus,

$$b = \mathrm{tr}(A_T^5) - a \tag{13.17}$$

where A_T is the adjacency matrix of $T_{3,1}$. Computing A_T^5 explicitly we find that $\mathrm{tr}(A_T^5) = 40$ and, again,

$$b = 40 - 30 = 10.$$

Finally, to find c, note that for every node in C_5 there is one CW of length five in a clockwise direction and another anticlockwise, e.g. $i \to j \to k \to l \to m \to i$ and $i \to m \to l \to k \to j \to i$. Thus, $c = 10$. Finally,

$$|C_5| = \frac{1}{10}\left(\mu_5 - 30|C_3| - 10|T_{3,1}|\right). \tag{13.18}$$

13.2.5 Formulae for counting small subgraphs

Formulae for a number of simple subgraphs can be derived using very similar techniques to the ones we have encountered so far. The results are summarized over the next few pages.

$F1$

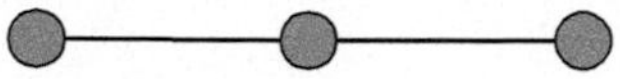

$$|F_1| = \frac{1}{2} \sum_i k_i(k_i - 1)$$

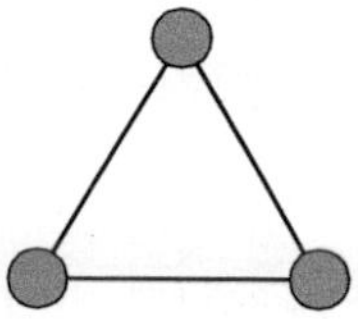

$F2$

$$|F_2| = \frac{1}{6} \mathrm{tr}(A^3)$$

$F3$

$$|F_3| = \sum_{(i,j) \in E} (k_i - 1)(k_j - 1) - 3|F_2|$$

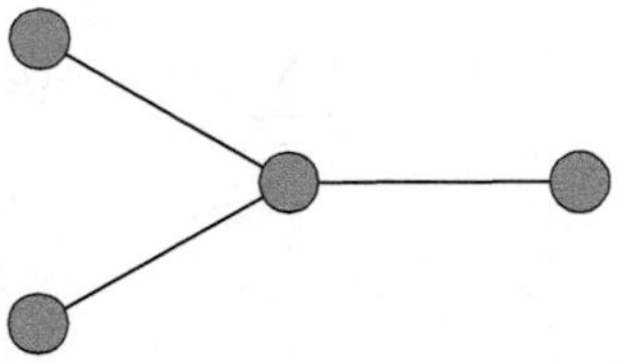

$F4$

$$|F_4| = \frac{1}{6} \sum_i k_i(k_i - 1)(k_i - 2)$$

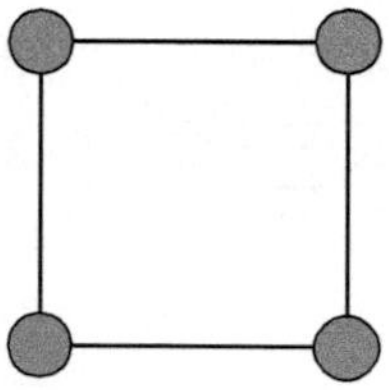

$F5$

$$|F_5| = \frac{1}{8}\left(\mathrm{tr}(A^4) - 4|F_1| - 2m\right)$$

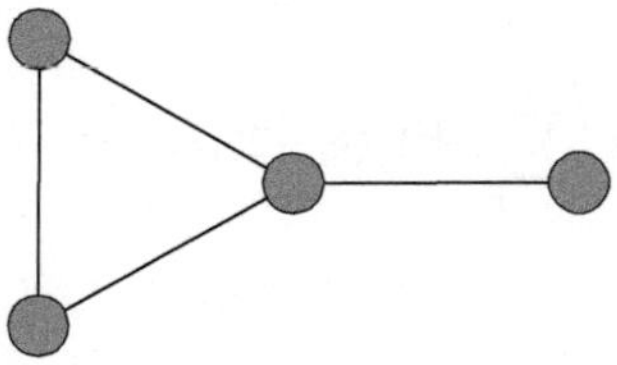

$F6$

$$|F_6| = \sum_{k_i > 2} t_i(k_i - 2)$$

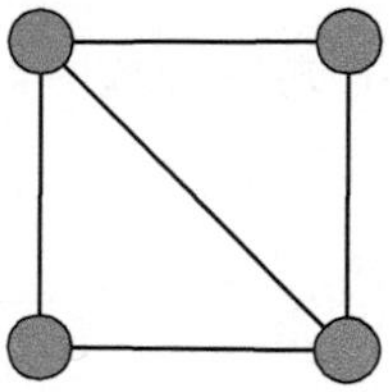

$F7$

$$|F_7| = \frac{1}{4} \sum_{i,j} \left((A^2)_{ij} A_{ij}\right)\left((A^2)_{ij} \cdot A_{ij} - 1\right)$$

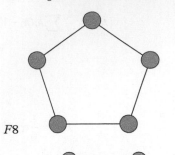

$F8$

$$|F_8| = \frac{1}{10}\left(\text{tr}(A^5) - 30|F_2| - 10|F_6|\right)$$

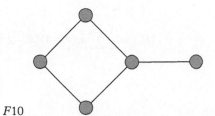

$F9$

$$|F_9| = \frac{1}{2}\sum_{k_i \geq 4} t_i(k_i - 2)(k_i - 3)$$

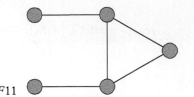

$F10$

$$|F_{10}| = \frac{1}{2}\sum_i (k_i - 2) \times \sum_{i,j}\binom{(A^2)_{ij}}{2} - 2|F_7|$$

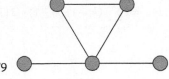

$F11$

$$|F_{11}| = \sum_{(i,j)\in E}(A^2)_{ij}(k_i - 2)(k_j - 2) - 2|F_7|$$

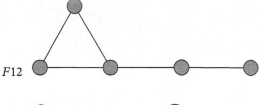

$F12$

$$|F_{12}| = \sum_i t_i\left(\sum_{i\neq j}(A^2)_{ij}\right) - 6|F_2| - 2|F_6| - 4|F_7|$$

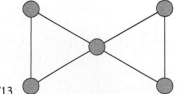

$F13$

$$|F_{13}| = \frac{1}{2}\sum_i t_i(t_i - 1) - 2|F_7|$$

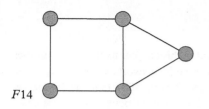

$F14$

$$|F_{14}| = \sum_{(i,j)\in E} (A^3)_{ij}(A^2)_{ij} - 9|F_2| - 2|F_6| - 4|F_7|$$

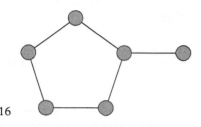

$F15$

$$|F_{15}| = \frac{1}{12}\Big(\mathrm{tr}(A^6) - 2m - 12|F_1| - 24|F_2| - 6|F_3| \\ -12|F_4| - 48|F_5| - 36|F_7| - 12|F_{10}| - 24|F_{13}|\Big)$$

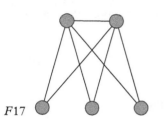

$F16$

$$|F_{16}| = \frac{1}{2}\sum_i (k_i - 2)B_i - 2|F_{14}| \quad \text{where}$$
$$B_i = (A^5)_{ii} - 20t_i - 8t_i(k_i - 2) - 2\sum_{(i,j)\in E}(A^2)_{ij}(k_j - 2) - 2\sum_{(i,j)\in E}\big(t_j - (A^2)_{ij}\big)$$

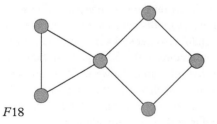

$F17$

$$|F_{17}| = \sum_{(i,j)\in E} \binom{(A^2)_{ij}}{3}$$

$F18$

$$|F_{18}| = \sum_i t_i \cdot \sum_{i\neq j}\binom{(A^2)_{ij}}{2} - 6|F_7| - 2|F_{14}| - 6|F_{17}|$$

13.3   Network motifs

We can use the techniques we have developed in this chapter to count the number of occurrences of a given fragment in a real-world network. Certain fragments arise inevitably through network connectivity. It is possible that the frequency with which they appear is similar to equivalent random networks. In this case, we cannot use the abundance of a fragment to explain any evolutionary mechanism giving rise to the structure of that network. However, if a given fragment appears more frequently than expected we can infer that there is some structural or functional reasons for the over expression. This is precisely the concept of a *network motif*. A subgraph is considered a network motif if the probability P of it appearing in a random network an equal or greater number of times than in the real-world network is lower than a certain cut-off value, which is generally taken to be $P_c = 0.01$.

In order to quantify the statistical significance of a given motif we use the Z-score which, for a given subgraph i, is defined as

$$Z_i = \frac{N_i^{real} - \langle N_i^{random} \rangle}{\sigma_i^{random}}, \qquad (13.19)$$

where N_i^{real} is the number of times the subgraph i appears in the real network, $\langle N_i^{random} \rangle$ and σ_i^{random} are the average and standard deviation of the number of times that i appears in an ensemble of random networks, respectively. Similarly, the relative abundance of a given fragment can be estimated using the statistic:

$$\alpha_i = \frac{N_i^{real} - \langle N_i^{random} \rangle}{N_i^{real} + \langle N_i^{random} \rangle}. \qquad (13.20)$$

13.3.1   Motifs in directed networks

Motifs in directed networks are simply directed subgraphs which appear more frequently in the real-world network than in its random counterpart. The situation is more complex because the number of motifs with the same number of nodes is significantly larger than for the undirected networks (for instance, there are seven directed triangles versus only one undirected); and in general there are no analytical tools for counting such directed subgraphs. However, there are several computational approaches that allow the calculation of the number of small directed subgraphs and directed motifs in networks. In Figure 13.8 we illustrate some examples of directed triangles found in real-world networks as motifs.

A characteristic feature of network motifs is that they are network-specific. That is, what is a motif for one is not necessarily a motif for another. However, a family of networks can be identified if they share the same series of motifs. One can characterize this by generating a vector whose ith entry gives the importance

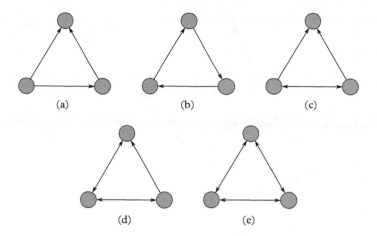

Figure 13.8 *Motifs in directed networks (a) Feed-forward loop (neurons) (b) Three-node feedback loop (Electronic circuits) (c) Up-linked mutual dyad (d) Feedback with mutual dyads (e) Fully connected triad. Dyads are typical in the WWW*

of the ith motif with respect to the other motifs in the network. The resulting component of the *significance profile* vector is given by

$$SP_i = \frac{Z_i}{\sqrt{\sum_i Z_j^2}}.$$

$$(13.21)$$

13.3.2 Motifs in undirected networks

In Figure 13.9 we illustrate the relative abundance of 17 of the small subgraphs we have discussed for six complex networks representing different systems in the real-world. The average is taken over random networks whose nodes have the same degrees as the real ones. It can be seen that there are a few fragments which are over-represented in some networks while other fragments are under-represented. Fragments which appear less frequently in a real-world network than is expected in an analogous random one are called *anti-motifs*.

Problem 13.3
The connected component of the protein–protein interaction network of yeast has 2,224 nodes and 6,609 links. It has been found computationally that the number of triangles in that network is 3,530. Determine the relative abundance of this fragment in order to see whether it is a motif in this network.

We use the formula

$$\alpha_i = \frac{t_i^{real} - \langle t_i^{random} \rangle}{t_i^{real} + \langle t_i^{random} \rangle},$$

where we know that $t_i^{real} = 3,530$. We have to estimate $\langle t_i^{random} \rangle$. Let us consider Erdös-Rényi random networks with 2,224 nodes and 6,609 links for which

$$p = \frac{2m}{n(n-1)} = 0.00267.$$

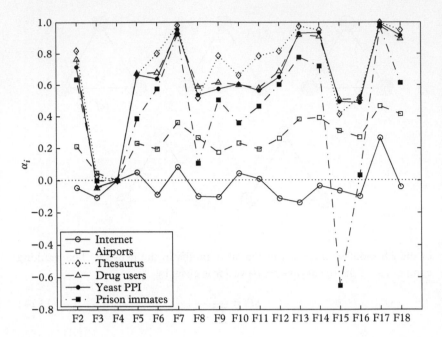

Figure 13.9 *Motifs and anti-motifs in undirected networks*

We also know that for large n, $\lambda_1 \to pn$ and all the other eigenvalues are negligible so we use the approximation

$$\langle t_i^{random} \rangle = \frac{1}{6} \sum_{j=1}^{n} \lambda_j^3 \approx \frac{\lambda_1^3}{6} = \frac{(np)^3}{6} .$$

Thus, $\langle t_i^{random} \rangle \approx 35$. This estimate is very good indeed. For instance, the average number of triangles in 100 realizations of an ER network is $\langle t_i^{random} \rangle = 35.4 \pm 6.1$. Using the value of $\langle t_i^{random} \rangle \approx 35$ we obtain $\alpha_i = 0.98$, which is very close to one. We conclude that the number of triangles in the yeast PPI is significantly larger than expected by chance, and we can consider it as a network motif.

FURTHER READING

Alon, N., Yuster, R., and Zwick, U., Finding and counting given length cycles, Algorithmica **17**:209–223, 1997.

Milo, R. et al., Network motifs: Simple building blocks of complex networks, Science **298**:824–827, 2002.

Milo, R. et al., Superfamilies of evolved and designed networks, Science **303**:1538–1542, 2004.

Classical Node Centrality

14

In this chapter

The concept of node centrality is motivated and introduced. Some properties of the degree of a node are analysed along with extensions to consider non-nearest neighbours. Two centralities based on shortest paths on the network are defined—the closeness and betweenness centrality—and differences between them are described. We finish this chapter with a few problems to illustrate how to find analytical expression for these centralities in certain classes of networks.

14.1 Motivation

The notion of centrality of a node first arose in the context of social sciences and is used in the determination of the most 'important' nodes in a network. There are a number of characteristics, not necessarily correlated, which can be used in determining the importance of a node. These include its ability to communicate directly with other nodes, its closeness to many other nodes, and its indispensability to act as a communicator between different parts of a network.

Considering each of these characteristics in turn leads to different centrality measures. In this chapter we study such measures and illustrate the different qualities of a network that they can highlight.

14.2 Degree centrality

The degree centrality simply corresponds to degree, and clearly measures the ability of a node to communicate directly with others. As we have seen, the degree of node i in a simple network G is defined using its adjacency matrix, A, as

$$k_i = \sum_{j=1}^{n} a_{ij} = (\mathbf{e}^T A)_i = (A\mathbf{e})_i. \tag{14.1}$$

So with degree centrality, i is more central than j if $k_i > k_j$. In a directed network, where in-degree and out-degree can be different, we can utilize degree to get two centrality measures, namely,

$$k_i^{in} = \sum_{i=1}^{n} a_{ji} = (e^T A)_i, \quad k_i^{out} = \sum_{j=1}^{n} a_{ij} = (Ae)_i. \tag{14.2}$$

The following are some elementary facts about the degree centrality. You are invited to prove these yourself.

1. $k_i = (A^2)_{ii}$.

2. $\sum_{i=1}^{n} k_i = 2m$, where m is the number of links.

3. $\sum_{i=1}^{n} k_i^{in} = \sum_{i=1}^{n} k_i^{out} = m$, where m is the number of links.

Example 14.1

Let us consider the network illustrated in Figure 14.1

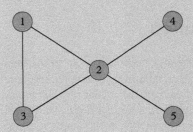

Figure 14.1 *A simple labelled network*

Since the adjacency matrix of the network is

$$A = \begin{bmatrix} 0 & 1 & 1 & 0 & 0 \\ 1 & 0 & 1 & 1 & 1 \\ 1 & 1 & 0 & 0 & 0 \\ 0 & 1 & 0 & 0 & 0 \\ 0 & 1 & 0 & 0 & 0 \end{bmatrix}$$

the node degree vector is

$$\mathbf{k} = Ae = \begin{bmatrix} 0 & 1 & 1 & 0 & 0 \\ 1 & 0 & 1 & 1 & 1 \\ 1 & 1 & 0 & 0 & 0 \\ 0 & 1 & 0 & 0 & 0 \\ 0 & 1 & 0 & 0 & 0 \end{bmatrix} \begin{bmatrix} 1 \\ 1 \\ 1 \\ 1 \\ 1 \end{bmatrix} = \begin{bmatrix} 2 \\ 4 \\ 2 \\ 1 \\ 1 \end{bmatrix}.$$

That is, the degrees of the nodes are: $k(1) = k(3) = 2$, $k(2) = 4$, $k(4) = k(5) = 1$, indicating that the most central node is 2.

Example 14.2

Let us consider the network displayed in Figure 14.2 together with its adjacency matrix

$$A = \begin{bmatrix} 0 & 1 & 0 & 1 & 1 & 0 \\ 1 & 0 & 0 & 0 & 0 & 0 \\ 0 & 1 & 0 & 1 & 0 & 0 \\ 0 & 0 & 0 & 0 & 0 & 0 \\ 0 & 0 & 0 & 1 & 0 & 0 \\ 1 & 0 & 0 & 0 & 0 & 0 \end{bmatrix}.$$

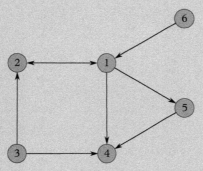

Figure 14.2 *A labelled directed network*

The in- and out-degree vectors are then obtained as follows:

$$\mathbf{k}^{in} = (\mathbf{e}^T A)^T = A^T \mathbf{e} = \begin{bmatrix} 0 & 1 & 0 & 1 & 1 & 0 \\ 1 & 0 & 0 & 0 & 0 & 0 \\ 0 & 1 & 0 & 1 & 0 & 0 \\ 0 & 0 & 0 & 0 & 0 & 0 \\ 0 & 0 & 0 & 1 & 0 & 0 \\ 1 & 0 & 0 & 0 & 0 & 0 \end{bmatrix}^T \begin{bmatrix} 1 \\ 1 \\ 1 \\ 1 \\ 1 \\ 1 \end{bmatrix} = \begin{bmatrix} 2 \\ 2 \\ 0 \\ 3 \\ 1 \\ 0 \end{bmatrix},$$

$$\mathbf{k}^{out} = A\mathbf{e} = \begin{bmatrix} 0 & 1 & 0 & 1 & 1 & 0 \\ 1 & 0 & 0 & 0 & 0 & 0 \\ 0 & 1 & 0 & 1 & 0 & 0 \\ 0 & 0 & 0 & 0 & 0 & 0 \\ 0 & 0 & 0 & 1 & 0 & 0 \\ 1 & 0 & 0 & 0 & 0 & 0 \end{bmatrix} \begin{bmatrix} 1 \\ 1 \\ 1 \\ 1 \\ 1 \\ 1 \end{bmatrix} = \begin{bmatrix} 3 \\ 1 \\ 2 \\ 0 \\ 1 \\ 1 \end{bmatrix}.$$

Nodes 3 and 6 are known as *sources* because their in-degrees are equal to zero but not their out-degrees. Node 4 is a *sink* because its out-degree is zero but not its in-degree. If both, the in- and out-degree are zero for a node, the node is isolated.

The most central node in sending information to its nearest neighbours is node 1 and in receiving information is node 4.

Example 14.3

Let us now consider a real-world network. It corresponds to the food web of St Martin island in the Caribbean, in which nodes represent species and food sources and the directed links indicate what eats what in the ecosystem. Here we represent the networks in Figure 14.3 by drawing the nodes as circles with radius proportional to the corresponding in-degree in (a) and out-degree in (b).

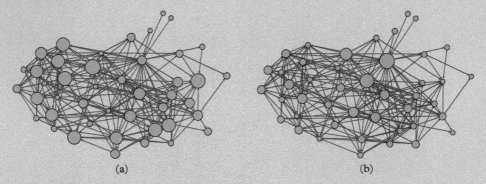

(a)   (b)

Figure 14.3 *Food webs in St Martin with nodes drawns as circles of radii proportional to (a) in-degree (b) out-degree*

The in- and out-degree vectors are calculated in exactly the same way as in the last example. In this case, every node has a label which corresponds to the identity of the species in question. In analysing this network according to the in- and out-degree, we can point out the following observations which are of relevance for the functioning of this ecosystem.

- Nodes with high out-degree are predators with a large variety of prey. Examples include the lizards *Anolis gingivinus* (the Anguilla Bank Anole) and *Anolis pogus*; and the birds the pearly-eyed thrasher and the yellow warbler.

- High in-degree nodes represent species and organic matter which are eaten by many others in this ecosystem, such as leaves, detritus, and insects such as aphids.

- In general, top predators are not predated by other species, thus having significantly higher out-degree than in-degree.

- The sources with zero in-degree are all birds: the pearly-eye thrasher, yellow warbler, kestrel, and grey kingbird.

- Highly predated species are not usually prolific predators, thus they have high in-degree but low out-degree.

- The sinks are all associated with plants or detritus.

14.3 Closeness centrality

The closeness centrality of a node characterizes how close this node is from the rest of the nodes. This closeness is measured in terms of the shortest path distance. The closeness of the node i in an undirected network G is defined as

$$CC(i) = \frac{n-1}{s(i)}, \qquad\qquad (14.3)$$

where the distance-sum $s(i)$ is calculated from the shortest path distances $d(i,j)$ as

$$s(i) = \sum_{j \in V(G)} d(i,j). \qquad\qquad (14.4)$$

In a directed network a node has in- and out-closeness centrality. The first corresponds to how close this node is to nodes it is receiving information from. The out-closeness centrality indicates how close the node is from those it is sending information to. In directed networks the shortest path is a pseudo-distance due to a possible lack of symmetry.

Example 14.4

Consider the network illustrated in Figure 14.4.

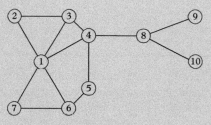

Figure 14.4 *A network where closeness centrality does not match degree centrality*

We start by constructing the distance matrix of this network, which is given by

$$D = \begin{bmatrix} 0 & 1 & 1 & 1 & 2 & 1 & 1 & 2 & 3 & 3 \\ 1 & 0 & 1 & 2 & 3 & 2 & 2 & 3 & 4 & 4 \\ 1 & 1 & 0 & 1 & 2 & 2 & 2 & 2 & 3 & 3 \\ 1 & 2 & 1 & 0 & 1 & 2 & 2 & 1 & 2 & 2 \\ 2 & 3 & 2 & 1 & 0 & 1 & 2 & 2 & 3 & 3 \\ 1 & 2 & 2 & 2 & 1 & 0 & 1 & 3 & 4 & 4 \\ 1 & 2 & 2 & 2 & 2 & 1 & 0 & 3 & 4 & 4 \\ 2 & 3 & 2 & 1 & 2 & 3 & 3 & 0 & 1 & 1 \\ 3 & 4 & 3 & 2 & 3 & 4 & 4 & 1 & 0 & 2 \\ 3 & 4 & 3 & 2 & 3 & 4 & 4 & 1 & 2 & 0 \end{bmatrix}$$

The vector of distance-sum of each node is then

$$\mathbf{s} = D\mathbf{e} = (\mathbf{e}^T D)^T = [\ 15 \quad 22 \quad 17 \quad 14 \quad 19 \quad 20 \quad 21 \quad 18 \quad 26 \quad 26\ ]^T$$

continued

Example 14.4 *continued*

And we use (14.3) to measure the closeness centrality of each node. For instance for node 1

$$CC(1) = \frac{9}{15} = 0.6.$$

The full vector of closeness centralities is

$$\mathbf{CC} = [0.600\ 0.409\ 0.529\ \mathbf{0.643}\ 0.474\ 0.450\ 0.428\ 0.500\ 0.346\ 0.346]^T,$$

indicating that the most central node is node 4. Notice that in this case the degree centrality identifies another node (namely 1) as the most important whereas in Figure 14.1 node 2 has both the highest degree and closeness centralities.

Example 14.5

We consider here the air transportation network of the USA, where the nodes represent the airports in the USA and the links represent the existence of at least one flight connecting the two airports. In Figure 14.5 we illustrate this network in which the nodes are represented as circles with radii proportional to the closeness centrality.

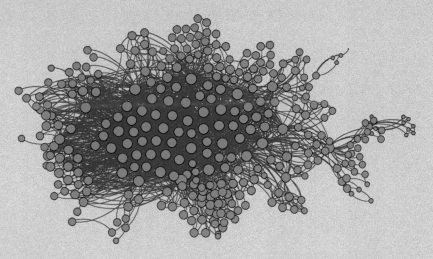

Figure 14.5 *A representation of the USA air transportation network in 1997*

The most central airports according to the closeness centrality are given in Table 14.1.

Table 14.1 *Central airports according to the closeness centrality.*

Airport	Closeness centrality × 100
Chicago O'Hare Intl	60.734
Dallas/Fort Worth Intl	55.444
Minneapolis-St Paul Intl	53.997
William B Hartsfield, Atlanta	53.560
San Francisco Intl	53.301
Lambert-St Louis Intl	52.875
Seattle-Tacoma Intl	52.623
Los Angeles Intl	52.456

The first four airports in this list (and the sixth) correspond to airports in the geographic centre area of continental USA. The other three are airports located on the west coast. The first group are important airports in connecting the East and West of the USA with an important traffic also between north and south of the continental USA. The second group represents airports with important connections between the main USA and Alaska, as well as overseas territories like Hawaii and other Pacific islands. The most highly ranked airports according to degree centrality are given in Table 14.2. Notice that the group of west coast airports is absent.

Table 14.2 *Highly ranked airports according to degree centrality.*

Airport	Degree centrality
Chicago O'Hare Intl	139
Dallas/Fort Worth Intl	118
William B Hartsfield, Atlanta	101
Pittsburgh Intl	94
Lambert-St Louis Intl	94
Charlotte/Douglas Intl	87
Stapleton Intl	85
Minneapolis-St Paul Intl	78

Problem 14.1

Let $CC(i)$ be the closeness centrality of the ith node in the path network P_{n-1} labelled $1 - 2 - 3 - 4 - \cdots - (n-1) - n$.

(a) Find a general expression for the closeness centrality of the ith node in P_{n-1} in terms of i and n only.

(b) Simplify the expressions found in (a) for the node(s) at the centre of the path P_{n-1} (for both odd and even values of n).

(c) Show that the closeness centrality of these central nodes is the largest in a path P_{n-1}.

The solution can be arrived at as follows

(a) We start by considering the sum of all the distances from one node to the rest of the nodes in the path.

Table 14.3 *The sum of all the distances from one node.*

Node	$\sum_{j \neq i} d_{ij}$
1	$1 + 2 + \cdots + n - 1$
2	$1 + 1 + 2 + \cdots + n - 2$
3	$2 + 1 + 1 + 2 + \cdots + n - 3$
$\vdots$	$\vdots$
i	$(i-1) + (i-2) + \cdots + 2 + 1 + 1 + 2 + \cdots + n - i$

It is important to notice here that for each node the sum of the distances corresponds to a 'right' sum, i.e. the sum of the distances of all nodes to the right of the node i and a 'left' sum, i.e. the sum of the distances of all nodes located to the left of i, $(i-1) + (i-2) + \cdots + 2 + 1$. These two sums are given, respectively, by

$$1 + 2 + \cdots + n - i = \frac{(n-i)(n-i+1)}{2}, \tag{14.5}$$

$$(i-1) + (i-2) + \cdots + 2 + 1 = \frac{(i-1)i}{2}. \tag{14.6}$$

Then, by substituting into the formula (14.3) we obtain

$$CC(i) = \frac{n-1}{\frac{(i-1)i}{2} + \frac{(n-i)(n-i+1)}{2}}, \tag{14.7}$$

which can be written as

$$CC(i) = \frac{2(n-1)}{(i-1)i + (n-i)(n-i+1)} .$$ (14.8)

(b) For a path with an odd number of nodes the central node is $i = \frac{n+1}{2}$.
By substitution into (14.8) we obtain

$$CC\left(\frac{n+1}{2}\right) = \frac{2(n-1)}{\left(\frac{n+1}{2}-1\right)\frac{n+1}{2} + \left(n-\frac{n+1}{2}\right)\left(n-\frac{n+1}{2}+1\right)} ,$$

(14.9)

which reduces to

$$CC\left(\frac{n+1}{2}\right) = \frac{4}{(n+1)} .$$ (14.10)

For a path with an even number of nodes the central nodes are $i = \frac{n}{2}$ and $i = \frac{n}{2} + 1$. Now,

$$CC\left(\frac{n}{2}\right) = \frac{2(n-1)}{\left(\frac{n}{2}-1\right)\frac{n}{2} + \left(n-\frac{n}{2}\right)\left(n-\frac{n}{2}+1\right)} ,$$ (14.11)

and in this case

$$CC\left(\frac{n}{2}\right) = \frac{4(n-1)}{n^2} .$$ (14.12)

We recover the same value when $i = \frac{n}{2} + 1$.

(c) Simply consider

$$CC(i+1) - CC(i) = \frac{2(n-1)}{(i+1)i + (n-i-1)(n-i)}$$
$$- \frac{2(n-1)}{i(i-1) + (n-i)(n-i+1)} .$$ (14.13)

Putting the right-hand side over a common denominator gives the numerator

$$4(n-1)(n-2i),$$ (14.14)

which is positive if $i < n/2$ and negative if $i > n/2$, so $CC(i)$ reaches its maximum value in the centre of the path.

14.4   Betweenness centrality

The betweenness centrality characterizes how important a node is in the communication between other pairs of nodes. That is, the betweenness of a node accounts for the proportion of information that passes through a given node in communications between other pairs of nodes in the network.

As for the closeness centrality, betweenness assumes that the information travels from one node to another through the shortest paths connecting those nodes. The betweenness of the node i in an undirected network G is defined as

$$BC(i) = \sum_j \sum_k \frac{\rho(j, i, k)}{\rho(j, k)}, \; i \neq j \neq k, \tag{14.15}$$

where $\rho(j, k)$ is the number of shortest paths connecting the node j to the node k, and $\rho(j, i, k)$ is the number of these shortest paths that pass through node i in the network.

If the network is directed, the term $\rho(j, i, k)$ refers to the number of directed paths from the node j to the node k that pass through the node i, and $\rho(j, k)$ to the total number of directed paths from the node j to the node k.

Example 14.6

We consider again the network used in Figure 14.4 and we explain how to obtain the betweenness centrality for the node labelled as one. For this, we construct Table 14.4 in which we give the number of shortest paths from any pair of nodes that pass through the node 1, $\rho(j, 1, k)$. We also report the total number of shortest paths from these pairs of nodes $\rho(j, k)$.

The betweenness centrality of the node 1 is simply the total sum of the terms in the last column of Table 14.4,

$$BC(1) = \sum_j \sum_k \frac{\rho(j, 1, k)}{\rho(j, k)} = 12.667.$$

Using a similar procedure, we obtain the betweenness centrality for each node:

$$\mathbf{BC} = [12.667 \; 0.000 \; 2.333 \; 20.167 \; 2.000 \; 1.833 \; 0.000 \; 15.000$$
$$0.000 \; 0.000]^T,$$

which indicates that the node 4 is the most central one, i.e. it is the most important in allowing communication between other pairs of nodes.

Table 14.4 *The number of shortest paths from any pair of nodes passing through 1, $\rho(j, 1.k)$*

(j,k)	$\rho(j,1,k)$	$\rho(j,k)$	$\dfrac{\rho(j,i,k)}{\rho(j,k)}$
2,4	1	2	1/2
2,5	2	3	2/3
2,6	1	1	1
2,7	1	1	1
2,8	1	2	1/2
2,9	1	2	1/2
2,10	1	2	1/2
3,6	1	1	1
3,7	1	1	1
4,6	1	2	1/2
4,7	1	1	1
6,8	1	2	1/2
6,9	1	2	1/2
6,10	1	2	1/2
7,8	1	1	1
7,9	1	1	1
7,10	1	1	1
			12.667

Example 14.7

In Figure 14.6 we illustrate the urban street network of the central part of Cordoba, Spain. The most central nodes according to the betweenness correspond to those street intersections which surround the central part of the city and connect it with the periphery.

continued

Example 14.7 *continued*

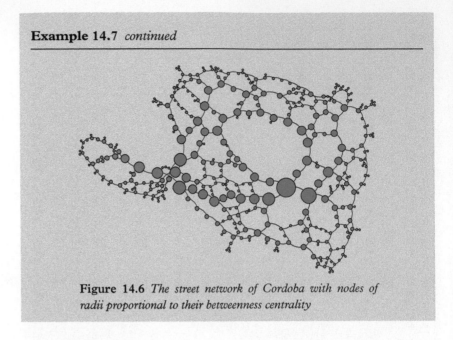

Figure 14.6 *The street network of Cordoba with nodes of radii proportional to their betweenness centrality*

Problem 14.2

Let G be a tree with $n = n_1 + n_2 + 1$ and the structure displayed in Figure 14.7. State conditions for the nodes labelled a, b, and c to have the largest value of betweenness centrality.

We start by considering the betweenness centrality of node a. Let us designate by V_1 and V_2 the two branches of the graph, the first containing n_1 and the second n_2 nodes.

Fact 1 Because the network is a tree, the number of shortest paths from p to q that pass through node k, $\rho(p, k, q)$, is the same as the number of shortest paths from p to q, $\rho(p, q)$. That is, $\rho(p, k, q) = \rho(p, q)$.

Fact 2 There are n_1 nodes in the branch V_1. Let us denote by i any node in this branch which is not a and by j any node in V_2 which is not c. There are $n_1 - 1$ shortest paths from nodes i to node b. That is,

$$\rho(i, a, b) = n_1 - 1. \tag{14.16}$$

Fact 3 We can easily calculate the number of paths from a node i to any node in the branch V_2 which go through node a. Because there are $n_1 - 1$ nodes of type i and n_2 nodes in the branch V_2 we have

$$\rho(i, a, V_2) = (n_1 - 1)n_2.$$

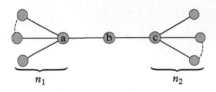

Figure 14.7 *A networked formed by joining two star networks together. Dashed lines indicate the existence of other equivalent nodes*

Fact 4 Any path going from a node denoted by i to another such node passes through the node a. Because there are $n_1 - 1$ nodes of the type i we have that the number of these paths is given by

$$\rho(i, a, i) = \binom{n_1 - 1}{2} = \frac{(n_1 - 1)(n_1 - 2)}{2}. \qquad (14.17)$$

Therefore the total number of paths containing the node a, and consequently its betweenness centrality, is

$$BC(a) = 2(n_1 - 1) + (n_1 - 1)(n_2 - 1) + \frac{(n_1 - 1)(n_1 - 2)}{2}$$
$$= \frac{(n_1 - 1)(2n_2 + n_1)}{2}.$$

In a similar way we obtain

$$BC(c) = \frac{(n_2 - 1)(2n_1 + n_2)}{2}. \qquad (14.18)$$

To calculate the betweenness centrality for node b we observe that every path from the n_1 nodes in branch V_1 to the n_2 nodes in branch V_2 passes through node b. Consequently,

$$BC(b) = n_1 n_2. \qquad (14.19)$$

Obviously, all nodes apart from $a, b,$ and c have zero betweenness centrality. We consider in turn the conditions for the three remaining nodes to be central.

In order for node a to have the maximum BC, the following conditions are necessary:

$$BC(a) > BC(c) \quad \text{and} \quad BC(a) > BC(b).$$

First,

$$BC(a) > BC(c) \Rightarrow \frac{(n_1)(2n_2 + n_1)}{2} > \frac{(n_2 - 1)(2n_1 + n_2)}{2}$$
$$\Rightarrow (n_1^2 + n_1) > (n_2^2 + n_2) \Rightarrow n_1 > n_2,$$

and

$$BC(a) > BC(b) \Rightarrow \frac{(n_1 - 1)(2n_2 + n_1)}{2} > n_1 n_2 \Rightarrow \frac{n_1(n_1 - 1)}{2} > n_2.$$

The second condition is fulfilled only if $n_1 \geq n_2$ and $n_1 > 3$. By combining both conditions we conclude that $BC(a)$ is the absolute maximum in the graph only in the cases when $n_1 > n_2$ and $n_1 > 3$.

By symmetry, $BC(c)$ is the absolute maximum if $n_2 > n_1$ and $n_2 > 3$.

Finally, for $BC(b)$ to be the absolute maximum we need

$$BC(b) > BC(a) \qquad\qquad BC(b) > BC(c)$$

$$n_1 n_2 > \frac{(n_1 - 1)(2n_2 + n_1)}{2} \quad \text{and} \quad n_1 n_2 > \frac{(n_2 - 1)(2n_1 + n_2)}{2}$$

$$n_2 > \frac{n_1(n_1 - 1)}{2} \qquad\qquad n_1 > \frac{n_2(n_2 - 1)}{2}$$

The two conditions are fulfilled simultaneously only if $n_1 = n_2 < 3$. That is, $BC(b)$ is the absolute maximum *only* when $n_1 = n_2 = 1$ or when $n_1 = n_2 = 2$, which correspond to P_2 and P_4, respectively (check this by yourself).

....

FURTHER READING

Borgatti, S.P., Centrality and network flow, Social Networks 27:55–71, 2005.

Borgatti, S.P. and Everett, M.G., A graph-theoretic perspective on centrality, Social Networks **28**:466–484, 2006.

Brandes, U. and Erlebach, T. (Eds.), *Network Analysis: Methodological Foundations*, Springer, 2005, Chapters 3–5.

Estrada, E., *The Structure of Complex Networks: Theory and Applications*, Oxford University Press, 2011, Chapter 7.

Wasserman, S. and Faust, K., *Social Network Analysis: Methods and Applications*, Cambridge University Press, 1994, Chapter 5.

Spectral Node Centrality

15

15.1 Motivation

Suppose we use a network to model a contagious disease amongst a population. Nodes represent individuals and edges represent potential routes of infection between these individuals. We illustrate a simple example in Figure 15.1. We focus on the nodes labelled 1 and 4 and ask which of them has the higher risk of contagion. Node 1 can be infected from nodes 2 and 3, while node 4 can be infected from 5 and 6. From this point of view it looks like both nodes are at the same level of risk. However, while 2 and 3 cannot be infected by any other node, nodes 5 and 6 can be infected from nodes 7 and 8, respectively. Thus, we can intuitively think that 4 is at a greater risk than 1 as a consequence of the chain of transmission of the disease. Local centrality measures like node degree do not account for a centrality that goes beyond the first nearest neighbours, so we need other kinds of measures to account for such effects. In this chapter we study these measures and illustrate the different qualities of a network that they can highlight.

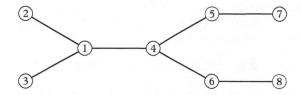

Figure 15.1 *A simple network. Nodes 1 and 4 are rivals for title of most central*

15.2 Katz centrality

The degree of the node i counts the number of walks of length one from i to every other node of the network. That is, $k_i = (A\mathbf{e})_i$. In 1953, Katz extended this idea to count not only the walks of length one, but those of any length starting at node i. Intuitively, we can reason that the closest neighbours have more influence over node i than more distant ones. Thus when combining walks of all lengths, one can introduce an attenuation factor so that more weight is given to shorter walks than to longer ones. This is precisely what Katz did and the Katz index is given by

$$K_i = \left[(\alpha^0 A^0 + \alpha A + \alpha^2 A^2 + \cdots + \alpha^k A^k + \cdots)\mathbf{e}\right]_i = \left[\sum_{k=0}^{\infty}(\alpha^k A^k)\mathbf{e}\right]_i. \quad (15.1)$$

The series in (15.1) is related to the resolvent function $(zI - A)^{-1}$. In particular, we saw in Example 12.4(iii) that the series converges so long as $\alpha < \rho(A)$ in which case

$$K_i = \left[(I - \alpha A)^{-1}\mathbf{e}\right]_i. \quad (15.2)$$

The Katz index can be expressed in terms of the eigenvalues and eigenvectors of the adjacency matrix. From the spectral decomposition $A = QDQ^T$ (see Chapter 5),

$$K_i = \sum_{l}\sum_{j} \mathbf{q}_j(i)\mathbf{q}_j(l)\frac{1}{1 - \alpha\lambda_j}. \quad (15.3)$$

When deriving his index, Katz ignored the contribution from $A^0 = I$ and instead used

$$\overline{K}_i = \left[\left((I - \alpha A)^{-1} - I\right)\mathbf{e}\right]_i. \quad (15.4)$$

While the values given by (15.2) and (15.4) are different, the rankings are exactly the same. We will generally use (15.2) because of the nice mathematical properties of the resolvent.

Example 15.1

We consider the network illustrated in Figure 15.1. The principal eigenvalue of the adjacency matrix for this network is $\lambda_1 = 2.1010$. With $\alpha = 0.3$ we obtain the vector of Katz centralities

$$K = [3.242 \quad 1.972 \quad 1.972 \quad \mathbf{3.524} \quad 2.591 \quad 2.591 \quad 1.777 \quad 1.777]^T.$$

Node 4 has the highest Katz index, followed by node 1 which accords with our intuition on the level of risk of each of these nodes in the network.

15.2.1   Katz centrality in directed networks

In directed networks we should consider the Katz centrality of a node in terms of the number of links going in and out from a node. This can be done by considering the indices

$$K_i^{out} = \left[(I - \alpha A)^{-1}\mathbf{e}\right]_i, \quad K_i^{in} = \left[\mathbf{e}^T(I - \alpha A)^{-1}\right]_i.$$

The second index can be considered as a measure of the 'prestige' of a node because it accounts for the importance that a node inherits from those that point to it.

Example 15.2

We measure the Katz indices of the nodes in the network illustrated in Figure 15.2.

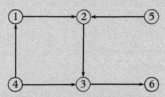

Figure 15.2 *A directed network. In- and out- centrality measures vary*

Using $\alpha = 0.5$ we obtain the Katz indices

$$K^{in} = \begin{bmatrix} 1.50 & 2.25 & 2.62 & 1.00 & 1.00 & 2.31 \end{bmatrix},$$
$$K^{out} = \begin{bmatrix} 1.88 & 1.75 & 1.50 & 2.69 & 1.88 & 1.00 \end{bmatrix}^T.$$

Notice that nodes 2 and 3 are each pointed to by two nodes. However, node 3 is more central because it is pointed to by nodes with greater centrality than those pointing to 2. In fact, node 6 is more central than node 2 because the only node pointing to it is the most important one in the network. On the other hand, out-Katz identifies node 4 as the most central one. It is the only node having out-degree of two.

15.3 Eigenvector centrality

Let us consider the following modification of the Katz index:

$$v = \left(\sum_{k=1}^{\infty} \alpha^{k-1} A^k \right) e = \left(\sum_{k=1}^{\infty} \alpha^{k-1} \sum_{j=1}^{n} q_j q_j^T \lambda_j^k \right) e = \left(\frac{1}{\alpha} \sum_{j=1}^{n} \sum_{k=1}^{\infty} (\alpha \lambda_j)^k q_j q_j^T \right) e$$

$$= \left(\frac{1}{\alpha} \sum_{j=1}^{n} \frac{1}{1-\alpha\lambda_j} q_j q_j^T \right) e.$$

Now, let the parameter α approach the inverse of the largest eigenvalue of the adjacency matrix from below, i.e. $\alpha \to 1/\lambda_1^-$. Then

$$\lim_{\alpha \to 1/\lambda_1^-} (1-\alpha\lambda_1) v = \lim_{\alpha \to 1/\lambda_1^-} \left(\frac{1}{\alpha} \sum_{j=1}^{n} \frac{(1-\alpha\lambda_1)v}{1-\alpha\lambda_j} q_j q_j^T \right) e = \left(\lambda_1 \sum_{i=1}^{n} q_1(i) \right) q_1 = \gamma q_1.$$

Thus the eigenvector associated with the largest eigenvalue of the adjacency matrix is a centrality measure conceptually similar to the Katz index. Accordingly, the eigenvector centrality of the node i is given by $q(i)$, the ith component of the principal eigenvector q_1 of A. Typically, we normalize q_1 so that its Euclidean length is one. By the Perron–Frobenius theorem we can choose q_1 so that all of its components are nonnegative.

Examples 15.3

(i) The eigenvector centralities for the nodes of the network in Figure 15.1 are

$$q_1 = [0.500 \quad 0.238 \quad 0.238 \quad \mathbf{0.574} \quad 0.354 \quad 0.354 \quad 0.168 \quad 0.168]^T.$$

Here again node 4 is the one with the highest centrality, followed by node 1. Node 4 is connected to nodes which are higher in centrality than the nodes to which node 1 is connected. High degree is not the only factor considered by this centrality measure. The most central nodes are generally connected to other highly central nodes.

(ii) Sometimes being connected to a few very important nodes make a node more central than being connected to many not so central ones. For instance, in Figure 15.3, node 4 is connected to only three other nodes, while 1 is connected to four. However, 4 is more central than 1 according to the eigenvector centrality because it is connected to two nodes with relatively high centrality while 1 is mainly connected to peripheral nodes. The vector of centralities is

$$q_1 = \begin{bmatrix} 0.408 & 0.167 & 0.167 & \mathbf{0.500} & 0.408 & 0.167 & 0.167 & 0.167 & 0.408 & 0.167 & 0.167 & 0.167 \end{bmatrix}^T.$$

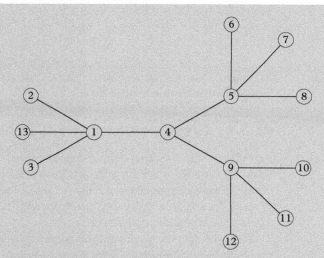

Figure 15.3 *A network highlighting the difference between degree and eigenvector centrality*

Problem 15.1

Let G be a simple connected network with n nodes and adjacency matrix A with spectral decomposition QDQ^T. Let $N_k(i)$ be the number of walks of length k starting at node i. Let

$$\mathbf{s}_k(i) = \frac{N_k(i)}{\sum_{j=1}^n N_k(j)}$$

be the ith element of the vector $\mathbf{s}_k$. Show that if G is not bipartite then there is a scalar α such that as $k \to \infty$, $\mathbf{s}_k \to \alpha \mathbf{q}_1$ almost surely. That is, the vector $\mathbf{s}_k$ will tend to rank nodes identically to eigenvector centrality.

Since $A^k = QD^kQ^T$,

$$\mathbf{s}_k(i) = \frac{\mathbf{e}_i^T A^k \mathbf{e}}{\mathbf{e}^T A^k \mathbf{e}} = \frac{\mathbf{e}_i^T QD^kQ^T \mathbf{e}}{\mathbf{e}^T QD^kQ^T \mathbf{e}} = \frac{\mathbf{q}_i^T D^k \mathbf{r}}{\mathbf{r}^T D^k \mathbf{r}} = \frac{\mathbf{q}_i^T \overline{D}^k \mathbf{r}}{\mathbf{r}^T \overline{D}^k \mathbf{r}},$$

where $\mathbf{r} = Q^T \mathbf{e}$ and $\overline{D} = D/\lambda_1$.

Since G is connected and not bipartite, $|\lambda_1| > |\lambda_j|$ for all $j > 1$ so $\overline{D}^k \to \mathbf{e}_1\mathbf{e}_1^T$ as $k \to \infty$.[1] We have established that

$$\mathbf{s}_k(i) \to \frac{\mathbf{q}_i^T \mathbf{e}_1\mathbf{e}_1^T \mathbf{r}}{\mathbf{r}^T \mathbf{e}_1\mathbf{e}_1^T \mathbf{r}} = \alpha \mathbf{q}_i(1)$$

[1] Note that $\mathbf{e}_1\mathbf{e}_1^T$ is a matrix whose only nonzero element is a 1 in the top left hand corner.

where $\alpha = 1/(\mathbf{e}_i^T \mathbf{r})$ and so

$$\lim_{k \to \infty} \mathbf{s}_k = \alpha \mathbf{q}_1,$$

as desired. Note that we require $\mathbf{e}_i^T \mathbf{r} \neq 0$ in this analysis, which is almost surely true for a network chosen at random.

15.3.1   Eigenvector centrality in directed networks

As with our other measures, we can define eigenvector centrality for directed networks. In this case we use the principal right and left eigenvectors of the adjacency matrix as the corresponding centrality vectors for the nodes in a directed network. If $A\mathbf{x} = \lambda_1 \mathbf{x}$ and $A^T\mathbf{y} = \lambda_1 \mathbf{y}$ then the elements of $\mathbf{x}$ and $\mathbf{y}$ give the right and left eigenvector centralities, respectively.

The right eigenvector centrality accounts for the importance of a node through the importance of nodes to which it points. It is an extension of the out-degree concept. On the other hand, the left eigenvector centrality accounts for the importance of a node by considering those nodes pointing towards a corresponding node and it is an extension of the in-degree centrality.

Example 15.4

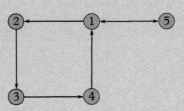

Figure 15.4 *A directed network highlighting the difference between left and right eigenvector centrality*

The left and right eigenvector centralities of the network in Figure 15.4 are

$$\mathbf{x} = \begin{bmatrix} 0.592 & 0.288 & 0.366 & 0.465 & 0.465 \end{bmatrix}^T, \quad \mathbf{y} = \begin{bmatrix} 0.592 & 0.465 & 0.366 & 0.288 & 0.465 \end{bmatrix}^T.$$

Notice the differences in the rankings of nodes 4 and 5. According to the right eigenvector, both nodes ranked as the second most central. They both point to the most central node of the network according to this criterion, node 1. However, according to the left eigenvector, while node 5 is still the second most important, node 4 has been relegated to the least central one. Node 5 is pointed to by the most central node, but node 4 is pointed to only by a node with low centrality.

15.3.2   PageRank centrality

When we carry out a search for a particular term, a search engine is likely to return thousands or millions of related web pages. A good search engine needs to make sure that pages that are most likely to match the query are promoted to the front of this list and centrality measures are a key tool in this process.

By viewing the World Wide Web (WWW) as a giant directed network whose nodes are pages and whose edges are the hyperlinks between them, search engines can attempt to rank pages according to centrality. While this is only one of the factors involved nowadays, much of the initial success of Google has been credited to their use of their own centrality measure, which they dubbed PageRank.

PageRank is closely related to eigenvector centrality and it explicitly measures the importance of a web page via the importance of other web pages pointing to it. In simple terms, the PageRank of a page is the sum of the PageRank centralities of all pages pointing into it.

The first step in computing PageRank is to manipulate the adjacency matrix, A. The WWW is an extremely complex network and is known to be disconnected. In order to apply familiar analytic tools, such as the Perron–Frobenius theorem, we need to make adjustments to A. In practice, the simplest approach is to artificially rewire nodes which have no outbound links so that they are connected to all the other nodes in the network. That is, we replace A with a new matrix H defined so

$$H_{ij} = \begin{cases} a_{ij}, & k_i^{out} > 0, \\ 1 & k_i^{out} = 0. \end{cases} \tag{15.5}$$

PageRank can then be motivated by considering what would happen if an internet surfer moved around the WWW from page to page by picking out-links uniformly at random. The insight that the developers of Google had was that the surfer is more likely to visit pages that have been deemed important in that they have in-links from other important pages. Using the theory of *Markov chains*, it can be shown that the relative frequency of page visits can be measured by the elements of the principal left eigenvector of the *stochastic matrix*

$$S = D^{-1}H,$$

where $D = \text{diag}(H\mathbf{e})$ is a diagonal matrix containing the out-degrees of the network with adjacency matrix H. Mathematically, PageRank is related to probability distributions so the vector of centralities is usually normalized to sum to one.

Of course, computing this eigenvector with a network as big as the WWW (which has billions of nodes) is a challenge in itself. For reasons of expediency, an additional parameter α (not to be confused with the one previously used for the Katz index) is introduced and rather than working with S we work with

$$P = \alpha S + \frac{1-\alpha}{n}\mathbf{e}\mathbf{e}^T. \tag{15.6}$$

The parameter is motivated by the suggestion that every so often, instead of following an out-link, our surfer teleports randomly to another page somewhere on the internet, preventing the user from getting stuck in a corner of the WWW. The value $\alpha = 0.85$ has been shown to work well in internet applications, but there is no reason whatsoever to use this same parameter value when PageRank is used on general complex networks.

Example 15.5

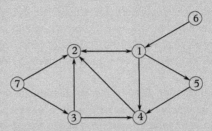

Figure 15.5 *A directed network illustrating PageRank centrality*

The normalized PageRank of the network in Figure 15.5 is

$$\mathbf{PG} = \begin{bmatrix} 0.301 & \mathbf{0.308} & 0.030 & 0.211 & 0.107 & 0.021 & 0.021 \end{bmatrix}^{T}$$

when $\alpha = 0.85$. Notice that node 1 has higher PageRank than node 4 due to its in-link from node 2. In this example, the rankings vary little as we change α.

15.4   Subgraph centrality

Katz centrality is computed from the entries of the matrix

$$K = \sum_{l=0}^{\infty} \alpha^{l} A^{l}.$$

We can easily generalize this idea and work with other weighted sums of the powers of the adjacency matrix, namely,

$$f(A) = \sum_{l=0}^{\infty} c_{l} A^{l}. \qquad (15.7)$$

The coefficients c_l are expected to ensure that the series is convergent; they should give more weight to small powers of the adjacency matrix than to the larger ones; and they should produce positive numbers for all $i \in V$.

Notice that if the first of the three requirements hold then (15.7) defines a matrix function and we can use theory introduced in Chapter 12. The diagonal entries, $f_i(A) = f(A)_{ii}$, are directly related to subgraphs in the network and the second requirement ensures that more weight is given to the smaller than to the bigger ones.

Example 15.6

Let us examine (15.7) when we truncate the series at $l = 5$ and select $c_l = \dfrac{1}{l!}$ to find an expression for $f_i(A)$.

Using information collected in Chapter 13 on enumerating small subgraphs we know that

$$(A^2)_{ii} = |F_1(i)|, \tag{15.8}$$

$$(A^3)_{ii} = 2|F_2(i)|, \tag{15.9}$$

$$(A^4)_{ii} = |F_1(i)| + |F_3(i)| + 2|F_4(i)| + 2|F_5(i)|, \tag{15.10}$$

$$(A^5)_{ii} = 10|F_2(i)| + 2|F_6(i)| + 2|F_7(i)| + 4|F_8(i)| + 2|F_9(i)|. \tag{15.11}$$

where the rooted fragments are illustrated in Figure 15.6. So,

$$
\begin{aligned}
f_i(A) = {} & (c_2 + c_4)|F_1(i)| + (2c_3 + 10c_5)|F_2(i)| + (c_4)|F_3(i)| + (2c_4)|F_4(i)| \\
& + (2c_4)|F_5(i)| + (2c_5)|F_6(i)| + (2c_5)|F_7(i)| + (4c_5)|F_8(i)| \\
& + (2c_5)|F_9(i)|.
\end{aligned}
\tag{15.12}
$$

By using $c_l = \dfrac{1}{l!}$ we get

$$
\begin{aligned}
f_i(A) = {} & \frac{13}{24}|F_1(i)| + \frac{5}{12}|F_2(i)| + \frac{1}{24}|F_3(i)| + \frac{1}{12}|F_4(i)| \\
& + \frac{1}{12}|F_5(i)| + \frac{1}{60}|F_6(i)| + \frac{1}{60}|F_7(i)| + \frac{1}{30}|F_8(i)| + \frac{1}{60}|F_9(i)|.
\end{aligned}
\tag{15.13}
$$

Clearly, the edges (and hence node degrees) are making the largest contribution to the centrality, followed by paths of length two, triangles, and so on.

continued

Example 15.6 *continued*

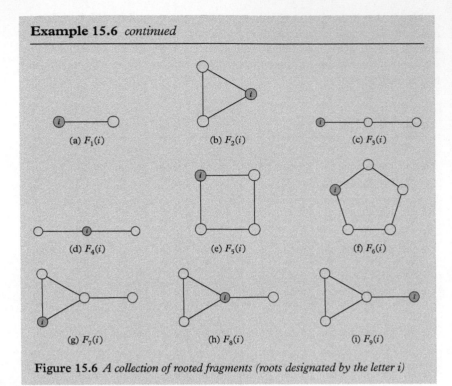

Figure 15.6 *A collection of rooted fragments (roots designated by the letter i)*

To define subgraph centrality we do not truncate (15.7) but work with the matrix functions which arise with particular choices of coefficients c_l. Some of the most well known are

$$EE_i = \left(\sum_{l=0}^{\infty} \frac{A^l}{l!}\right)_{ii} = \left(e^A\right)_{ii}, \tag{15.14}$$

$$EE_i^{odd} = \left(\sum_{l=0}^{\infty} \frac{A^{2l+1}}{(2l+1)!}\right)_{ii} = (\sinh(A))_{ii}, \tag{15.15}$$

$$EE_i^{even} = \left(\sum_{l=0}^{\infty} \frac{A^{2l}}{(2l)!}\right)_{ii} = (\cosh(A))_{ii}, \tag{15.16}$$

$$EE_i^{res} = \left(\sum_{l=0}^{\infty} \frac{A^l}{\alpha^l}\right)_{ii} = \left((I-\alpha A)^{-1}\right)_{ii}, \quad 0 < \alpha < 1/\lambda_1. \tag{15.17}$$

Notice that EE^{odd} and EE^{even} take into account only contributions from odd or even closed walks in the network, respectively. We will refer generically to EE as the subgraph centrality. Using the spectral decomposition of the adjacency matrix, these indices can be represented in terms of the eigenvalues and eigenvectors of the adjacency matrix as follows:

$$EE_i = \sum_{l=0}^{\infty} \mathbf{q}_l(i)^2 \exp(\lambda_l),$$

$$EE_i^{odd} = \sum_{l=0}^{\infty} \mathbf{q}_l(i)^2 \sinh(\lambda_l),$$

$$EE_i^{even} = \sum_{l=0}^{\infty} \mathbf{q}_l(i)^2 \cosh(\lambda_l),$$

$$EE_i^{res} = \sum_{l=0}^{\infty} \frac{\mathbf{q}_l(i)^2}{1-\alpha\lambda_l}, \quad 0 < \alpha < 1/\lambda_1.$$

Example 15.7

We compare some centrality measures for the network illustrated in Figure 15.7.

The regularity means that most centrality measures are unable to distinguish between nodes. The degree of each node is equal to six. Also, because the network is regular, $\mathbf{q}_1 = \mathbf{e}/3$ and the closeness and betweenness centralities are uniform with $CC(i) = 0.8$ and $BC(i) = 2$ for all $i \in V$. Observe also that each node is involved in ten triangles of the network.

The subgraph centrality, however, differentiates two groups of nodes $\{1,3,4,6,8\}$ with $EE_i = 45.65$ and $\{2,4,7,9\}$ with $EE_i = 45.70$. This indicates that the nodes in the second set participate in a larger number of small subgraphs than those in the first group. For instance, each node in the second group takes part in 45 squares versus 44 for the nodes in the first group.

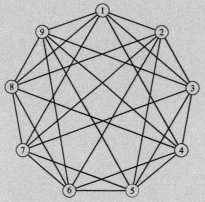

Figure 15.7 *A regular graph with common degree 6*

15.4.1   Subgraph centrality in directed networks

Subgraph centrality can be calculated for both directed and undirected networks using (15.14). Recall that a directed closed walk is a succession of directed links of the form $uv, vw, wx, \ldots, yz, zu$. This means that the subgraph centrality of a node in a directed network is $EE(i) > 1$ only if there is at least one closed walk that starts and returns to this node. Otherwise $EE(i) = 1$. The subgraph centrality of a node in a directed network indicates the 'returnability' of information to this node.

Example 15.8

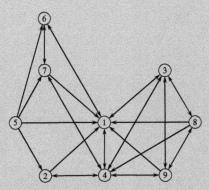

Figure 15.8 *A network representing the observed flow of votes from 2000–2013 between groups of countries in the Euro-vision song contest*

In the Eurovision song contest, countries vote for their favourite songs from other countries. We can represent these countries as nodes and the votes as directed edges. The aggregate voting over the 2000–2013 contests has been measured with links weighted according to the sum of votes between countries over the 14 years.[2] The countries can be grouped together according to their pattern of votes. Groups which vote in a similar way are represented by the directed network illustrated in Figure 15.8. The labels correspond to countries as follows.

1. Azerbaijan, Ukraine, Georgia, Russia, Armenia, Belarus, Poland, Bulgaria, Czech Republic.
2. Netherlands, Belgium.
3. Moldova, Romania, Italy, Israel.
4. Turkey, Boznia-Herzegovina, Macedonia, Albania, Serbia, Croatia, Slovenia, Austria, France, Montenegro.

[2] Details at tinyurl.com/oq9kpj7.

5. Ireland, United Kingdom, Malta.

6. Estonia, Lithuania, Latvia, Slovakia.

7. Iceland, Denmark, Sweden, Norway, Finland, Hungary.

8. Spain, Portugal, Germany, Andorra, Monaco, Switzerland.

9. Greece, Cyprus, San Marino.

The directed subgraph centrality for the groups of countries are

$$EE = \begin{bmatrix} 7.630 & 2.372 & 8.579 & \mathbf{12.044} & 1.000 & 2.950 & 3.553 & 5.431 & 5.431 \end{bmatrix}^{T}.$$

The highest 'returnabilities' of votes are obtained for groups 4, 3, and 1. The lowest returnability of votes is observed for group 5, which has no returnable votes at all, followed by group 2. Curiously, no countries from these two groups have won the contest in the last 14 years, while all the other groups (except group 3, which last won in 1998) have won the contest at least once in that time.

....

FURTHER READING

Langville, A.N. and Meyer, C.D., *Google's PageRank and Beyond: The Science of Search Engine Rankings*, Princeton University Press, 2006.

Estrada, E., *The Structure of Complex Networks: Theory and Applications*, Oxford University Press, 2011, Chapter 7.2.

Newmann, M.E.J., *Networks: An Introduction*, Oxford University Press, 2010, Chapter 7.

16 Quantum Physics Analogies

In this chapter

We introduce the basic principles and formalism of quantum mechanics. We study the quantum harmonic oscillator and introduce ladder operators. Then, we introduce the simplest model to deal with quantum (electronic) systems, the tight-binding model. We show that the Hamiltonian of the tight-binding model of a system represented by a network corresponds to the adjacency matrix of that network and its eigenvalues correspond to the energy levels of the system. We briefly introduce the Hubbard and Ising models.

16.1 Motivation

In Chapter 8 we used classical mechanics analogies to study networks. In a similar way, we can use quantum mechanics analogies. Quantum mechanics is the mechanics of the microworld. That is, the study of the mechanical properties of particles which are beyond the limits of our perception, such as electrons and photons. We remark here again that our aim is not simply to consider complex networks in which the entities represented by nodes behave quantum mechanically but to use quantum mechanics as a metaphor that allows us to interpret some of the mathematical concepts we use for studying networks in an amenable physical way. At the same time we aim to use elements of the arsenal of techniques and methods developed for studying quantum systems in the analysis of complex networks. Analogies are just that: analogies.

As we will see in this chapter, through the lens of quantum mechanics we can interpret the spectrum of the adjacency matrix of a network as the energy levels of a quantum system in which an electron is housed at each node of the network. This will prepare the terrain for more sophisticated analysis of networks in terms of how information is diffused through their nodes and links. And we will investigate other theoretical tools, such as the Ising model, which have applications in the analysis of social networks. Thus, when we apply these models to networks we will be equipped with a better understanding of the physical principles used in them.

16.2 Quantum mechanical analogies

We start this chapter with the postulates of quantum mechanics.

1. A quantum mechanical system is defined by a Hilbert space $\mathcal{H}$ containing complex-valued vectors $\psi(x)$ known as wave functions which describe the quantum state of moving particles. Wave functions are often normalized so that

$$\int_{-\infty}^{\infty} |\psi(x)|^2 dx = 1. \tag{16.1}$$

2. Every physical observable is represented by a linear differential operator which acts on $\psi(x)$ and is self-adjoint or Hermitian, i.e. for the operator $\hat{O}, \hat{O} = \hat{O}^*$. That is,

$$\int_{-\infty}^{\infty} \psi^*(x)(\hat{O}\phi)(x)\,dx = \int_{-\infty}^{\infty} (\hat{O}\psi)^*(x)\phi(x)\,dx, \tag{16.2}$$

for all square-integrable[1] wave functions ϕ and ψ.

3. In any measurement of the observable associated with the operator $\hat{O}$, the only values that will ever be observed are the eigenvalues λ, which satisfy the eigenvalue equation

$$\hat{O}\psi = \lambda\psi. \tag{16.3}$$

4. A complete description of the system is obtained from the normalized state vector ψ, such as through

$$\langle \hat{O} \rangle_\psi = \int_{-\infty}^{\infty} \psi^* \hat{O} \psi \, dx, \tag{16.4}$$

which represents the expected result of measuring the observable $\hat{O}$.

5. The time evolution of a state of a system is determined by the time-dependent Schrödinger equation

$$i\hbar \frac{d\psi}{dt} = \hat{\mathcal{H}}\psi, \tag{16.5}$$

where $\hat{\mathcal{H}}$ is the Hamiltonian or total energy operator. If the Hamiltonian is independent of time, the energy levels of the system are obtained from the eigenvalue equation

$$\hat{\mathcal{H}}\psi(r, t) = E\psi(r, t), \tag{16.6}$$

which then evolves in time according to

$$\psi(r, t) = e^{\frac{-itE}{\hbar}} \psi(r, 0). \tag{16.7}$$

[1] That is, $\int_{-\infty}^{\infty} |\psi(x)|^2 dx < \infty$.

It is common in quantum mechanics and its applications to network theory to use the so-called Dirac notation. Using this notation we can make the following substitutions:

(i) $\psi(x) \rightarrow |\psi\rangle$.

(ii) $\psi^*(x) \rightarrow \langle\psi|$.

(iii) $\psi(x) = c_1\psi_1(x) + c_2\psi_2(x) \rightarrow |\psi\rangle = c_1|\psi_1\rangle + c_2|\psi_2\rangle$.

(iv) $\int \phi^*(r)\psi(r)dr \rightarrow \langle\phi|\psi\rangle$.

(v) $\hat{O}\psi(r) \rightarrow \hat{O}|\psi\rangle = |\hat{O}\psi\rangle$.

(vi) $\int \phi^*(r)\hat{O}\psi(r)dr \rightarrow \langle\phi|\hat{O}\psi\rangle$.

To warm up, we will start by considering a simple harmonic oscillator (SHO). Recall from Chapter 8 that the Hamiltonian for this system can be written

$$\mathcal{H}(x,p) = \frac{p^2}{2m} + \frac{1}{2}kx^2 = \frac{p^2}{2m} + \frac{1}{2}m\omega^2 x^2. \tag{16.8}$$

In our new notation, quantizing the system gives

$$\hat{\mathcal{H}} = \frac{1}{2m}\hat{p}_x^2 + \frac{1}{2}m\omega^2\hat{x}^2. \tag{16.9}$$

For the momentum we have

$$\hat{p}_x = -i\hbar\frac{\partial}{\partial x}. \tag{16.10}$$

Let us first see what happens if we apply these two operators in a different order to a given function $\phi(x)$. That is,

$$\hat{x}\hat{p}_x\phi(x) = -i\hbar x\frac{\partial\phi(x)}{\partial x}, \quad \hat{p}_x\hat{x}\phi(x) = -i\hbar\frac{\partial}{\partial x}[x\phi(x)].$$

We obtain one of the fundamental results of quantum mechanics

$$\left(\hat{x}\hat{p}_x - \hat{p}_x\hat{x}\right)\phi(x) = -i\hbar\left[x\frac{\partial\phi(x)}{\partial x} - \phi(x)\frac{\partial x}{\partial x} - x\frac{\partial\phi(x)}{\partial x}\right] = i\hbar\phi(x). \tag{16.11}$$

That is, the momentum and the coordinates do not commute. A common alternative representation of (16.11) is $[\hat{x}, \hat{p}_x] = i\hbar$.

Since

$$\hat{p}_x^2 = \left(-i\hbar\frac{\partial}{\partial x}\right)\left(-i\hbar\frac{\partial}{\partial x}\right) = -\hbar^2\frac{\partial^2}{\partial x^2}, \tag{16.12}$$

we can rewrite (16.9) to express the Hamiltonian operator for the SHO as

$$\hat{\mathcal{H}} = \frac{\hbar^2}{2m}\frac{d^2}{dx^2} + \frac{1}{2}m\omega^2\hat{x}^2. \tag{16.13}$$

From (16.6),

$$-\frac{\hbar^2}{2m}\frac{d^2\psi}{dx^2} + \frac{1}{2}m\omega^2 x^2\psi = E\psi. \tag{16.14}$$

Letting $u = \sqrt{\dfrac{m\omega}{\hbar}}\,x$ and $\varepsilon = \dfrac{2E}{\hbar\omega}$, (16.14) becomes

$$\frac{d^2\psi}{du^2} + (u - \varepsilon^2)\psi = 0. \tag{16.15}$$

Solutions of this second order differential equation can be written as

$$\psi_j(u) = H_j(u)e^{-u^2/2}, \tag{16.16}$$

where $H_j(z)$ is the Hermite polynomial (see Further reading)

$$H_j(z) = (-1)^j\, e^{z^2}\frac{d^j}{dz^j}\left(e^{-z^2}\right). \tag{16.17}$$

Thus

$$\psi_j(x) = \frac{1}{\sqrt{2^n\,n!}}\left(\frac{m\omega}{\pi\hbar}\right)^{1/4} H_j\left(\sqrt{\frac{m\omega}{\hbar}}\,x\right)e^{-m\omega x^2/2\hbar}. \tag{16.18}$$

Applying the series solution to the Schrödinger equation we obtain the values of the energy of the quantum SHO

$$E_j = \hbar\omega\left(j + \frac{1}{2}\right), \quad j = 0, 1, 2, \ldots. \tag{16.19}$$

This is notably different from the classical SHO because now the energy can take only certain discrete values, i.e. it is quantized. Indeed the first energy levels of the SHO are:

$$E_0 = \frac{1}{2}\hbar\omega, \quad E_1 = \frac{3}{2}\hbar\omega, \quad E_2 = \frac{5}{2}\hbar\omega, \ldots.$$

A useful technique in solving the quantum SHO is to use the so-called ladder operators. The so-called annihilation (lowering) operator $\hat{c}$ and the creation (rising) operator $\hat{c}^{\dagger}$ are defined as

$$\hat{c} = \sqrt{\frac{m\omega}{2\hbar}}\left(\hat{x} + \frac{i\hat{p}}{m\omega}\right), \tag{16.20}$$

$$\hat{c}^{\dagger} = \sqrt{\frac{m\omega}{2\hbar}}\left(\hat{x} - \frac{i\hat{p}}{m\omega}\right). \tag{16.21}$$

These satisfy the commutation relation $[\hat{c}, \hat{c}^\dagger] = 1$. Additionally

$$[\hat{c}, \hat{\mathcal{H}}] = \hbar\omega\hat{c}, \quad [\hat{c}^\dagger, \hat{\mathcal{H}}] = -\hbar\omega\hat{c}. \tag{16.22}$$

The Hamiltonian of the quantum SHO can be written in terms of the ladder operators as

$$\hat{\mathcal{H}} = -\hbar\omega\left(\hat{c}^\dagger\hat{c} + \frac{1}{2}\right) = -\hbar\omega\left(\hat{\mathcal{N}} + \frac{1}{2}\right), \tag{16.23}$$

where $\hat{\mathcal{N}}$ is the so-called number operator.

Problem 16.1
Show that $\hat{c}$ lowers the energy of a state by an amount $\hbar\omega$ and that $\hat{c}^\dagger$ raises the energy by the same amount.
From (16.22), $\hat{\mathcal{H}}\hat{c} = \hat{c}\,\hat{\mathcal{H}} - \hbar\omega\hat{c}$. Now consider the effect of $\hat{\mathcal{H}}$ on the action of applying the annihilation operator to a state of the system, $\hat{c}|j\rangle$. That is,

$$\hat{\mathcal{H}}\left(\hat{c}|j\rangle\right) = \left(\hat{c}\hat{\mathcal{H}} - \hbar\omega\hat{c}\right)|j\rangle = (E_j - \hbar\omega)\left(\hat{c}|j\rangle\right).$$

Then, the operator $\hat{c}$ has lowered the energy E_j of $|j\rangle$ by $\hbar\omega$.
Similarly $\hat{\mathcal{H}}\hat{c}^\dagger = \hat{c}^\dagger\hat{\mathcal{H}} + \hbar\omega\hat{c}^\dagger$ and so

$$\hat{\mathcal{H}}\left(\hat{c}^\dagger|j\rangle\right) = \left(\hat{c}^\dagger\hat{\mathcal{H}} + \hbar\omega\hat{c}^\dagger\right)|j\rangle = (E_j + \hbar\omega)\left(\hat{c}^\dagger|j\rangle\right),$$

which indicates that the operator $\hat{c}^\dagger$ has raised the energy E_j of $|j\rangle$ by $\hbar\omega$.

16.3 Tight-binding models

To use our quantum models on networks we assume that we place an electron at each node of a network. Because the nodes are so heavy in comparison with the electrons we can assume that the properties of the system are mainly determined by the dynamics of these electrons. Here, the electrons can play the role of information which can be transferred between the nodes. This analogy allows us to use the techniques of condensed matter physics in which the properties of the solid state and of certain molecular systems are determined by considering a Hamiltonian of the form

$$\hat{\mathcal{H}} = \sum_{j=1}^{n}\left[-\frac{\hbar^2}{2m}\hat{p}_j^2 + \sum_{k\neq j}V(r_j - r_k) + U(r_j)\right], \tag{16.24}$$

where $V(r_j - r_k)$ is the potential describing the interactions between electrons and $U(r_j)$ is an external potential which we will assume is zero.

The electron has a property which is unknown in classical physics, called the spin. It is an intrinsic form of angular momentum and mathematically it can be described by a state in the Hilbert space

$$\alpha\,|+\rangle + \beta\,|-\rangle, \tag{16.25}$$

which is spanned by the basis vectors $|\pm\rangle$. Using the ladder operators previously introduced, the Hamiltonian (16.24) can be written as

$$\hat{\mathcal{H}} = -\sum_{ij} t_{ij} \hat{c}_i^\dagger \hat{c}_j + \frac{1}{2} \sum_{ijkl} V_{ijkl} \hat{c}_i^\dagger \hat{c}_k^\dagger \hat{c}_l \hat{c}_j, \tag{16.26}$$

where t_{ij} and V_{ijkl} are integrals which control the hopping of an electron from one site to another and the interaction between electrons, respectively. We can further simplify our Hamiltonian if we suppose that the electrons do not interact with each other, so all the V_{ijkl} equal zero. This method, which is known as the tight-binding approach or the Hückel molecular orbital method is very useful to calculate the properties of solids and molecules, like graphene. The Hamiltonian of the system becomes

$$\hat{\mathcal{H}}_{tb} = -\sum_{ij} t_{ij} \hat{c}_{i\rho}^\dagger \hat{c}_{i\rho}, \tag{16.27}$$

where $\hat{c}_{i\rho}$ creates (and $\hat{c}_{i\rho}^\dagger$ annihilates) an electron with spin ρ at the node i. We can now separate the in-site energy α_i from the transfer energy β_{ij} and write the Hamiltonian as

$$\hat{\mathcal{H}}_{tb} = \sum_{ij} \alpha_i \hat{c}_{i\rho}^\dagger \hat{c}_{i\rho} + \sum_{(ij)\rho} \beta_{ij} \hat{c}_{i\rho}^\dagger \hat{c}_{i\rho}, \tag{16.28}$$

where the second sum is carried out over all pairs of nearest-neighbours. Consequently, in a network with n nodes, the Hamiltonian (16.28) is reduced to an $n \times n$ matrix,

$$\hat{\mathcal{H}}_{ij} = \begin{cases} \alpha_i, & i = j, \\ \beta_{ij}, & i \text{ is connected to } j, \\ 0, & \text{otherwise.} \end{cases} \tag{16.29}$$

Assuming a homogeneous geometrical and electronic configuration, it is appropriate to give fixed values to the α_i and the β_{ij}. Typically, $\alpha_i = \alpha$ is a Fermi energy and $\beta_{ij} = \beta$ is fixed at -2.70eV for all pairs of connected nodes. Thus,

$$\hat{\mathcal{H}} = \alpha I + \beta A, \tag{16.30}$$

where I is the identity matrix, and A is the adjacency matrix of the graph representing the electronic system. The energy levels of the system are simply given by the eigenvalues of the adjacency matrix of the network:

$$E_j = \alpha + \beta\lambda_j. \tag{16.31}$$

Notice that since $\beta < 0$ we can interpret the eigenvalues of the adjacency matrix of a network as the negative of the energy levels of a tight-binding system, as described in this section. This will be very useful when we introduce statistical mechanics concepts for networks. For each energy level the molecular orbitals are constructed as linear combinations of the corresponding atomic orbitals for all the atoms in the system. That is,

$$\boldsymbol{\psi}_j = \sum_i c_j(i) \mathbf{q}_j(i), \tag{16.32}$$

where $\mathbf{q}_j(i)$ is the ith entry of the jth eigenvector of the adjacency matrix and $c_j(i)$ are coefficients of a linear combination.

Example 16.1

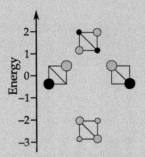

Figure 16.1 *A representation of eigenvalues and eigenvectors as energy levels and orbitals*

In Figure 16.1 we represent the eigenvalues of the adjacency matrix of a network as the negative energy levels. The corresponding entries of the eigenvectors are then represented as the 'network orbitals' (the equivalent of molecular orbitals in tight-binding theory). The node contributions (atomic orbitals) to the network orbital for the lowest energy level (principal eigenvalue) can be used as a measure of the centrality of the nodes in a network.

16.4 Some specific quantum-mechanical systems

16.4.1 The Hubbard model

The Hubbard model is an extension of the tight-binding Hamiltonian in which we introduce electron–electron interactions. To keep things simple, we allow on-site interactions only. That is, we consider one electron per site and $V_{ijkl} \neq 0$ in (16.26) if and only if i, j, k, and l all refer to the same node. In this case the Hamiltonian is written as

$$\hat{\mathcal{H}} = -t \sum_{i,j,\sigma} A_{ij} \hat{c}_{i\sigma}^\dagger \hat{c}_{j\sigma} + U \sum_i \hat{c}_{i\uparrow}^\dagger \hat{c}_{i\uparrow} \hat{c}_{i\downarrow}^\dagger \hat{c}_{i\downarrow}, \qquad (16.33)$$

where t is the hopping parameter and $U > 0$ indicates that the electrons repel each other.

Notice that if there is no electron–electron repulsion ($U = 0$), we recover the tight-binding Hamiltonian studied in the previous section.

16.4.2 The Ising model

Again, we consider a network in which we place an electron per node. Each electron can have either spin up (+) or down (–), as illustrated in Figure 16.2.

In the Ising model, we make the following assumptions.

(i)   Two spins interact only if they are located in nearest neighbour nodes.

(ii) The interaction between every pair of spins has the same strength.

(iii) The energy of the system decreases with the interaction between two identical spins and increases otherwise.

(iv) Each spin can interact with an external magnetic field H.

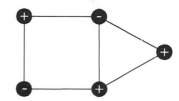

Figure 16.2 *A network with nodes signed according to spin up (+) or down (–)*

By combining these four assumptions we end up with a simplified Hamiltonian for the spin system, namely,

$$\hat{\mathcal{H}} = -\mathcal{J} \sum_{(i,j) \in E} \sigma_i \sigma_j - H \sum_i \sigma_i, \qquad (16.34)$$

where $\sigma_i = \pm 1$ represents the direction of the spin. If $\mathcal{J} > 0$, the interaction is called ferromagnetic and if $\mathcal{J} < 0$ it is antiferromagnetic. Usually the system is simplified by assuming no external magnetic field, hence,

$$\hat{\mathcal{H}} = -\mathcal{J} \sum_{(i,j) \in E} \sigma_i \sigma_j. \qquad (16.35)$$

Often we let $\mathcal{J} = \beta = (k_B T)^{-1}$ be the inverse temperature of the system, where k_B is the Boltzmann constant (more details in Chapter 20). Then, at low temperature, configurations in which most spins are aligned have lower energy. It is easy to imagine some potential applications of the Ising model in studying complex networks. If we consider a social network in which the nodes represent people and the links their social interactions, the spin can represent a vote in favour or against a certain statement. One can use the model to investigate whether the local alignment of voting or opinions among the nodes can generate a global state of consensus in the whole network.

....

FURTHER READING

McMahon, D., *Quantum Mechanics Demystified*, McGraw Hill Education, 2013.
Holznet, S., *Quantum Physics for Dummies*, John Wiley and Sons, 2013.
Cipra B.A., An introduction to the Ising model, American Mathematical Monthly 94:937–959, 1987.

Global Properties of Networks I

17

In this chapter

We study the correlation between the degrees of the nodes connected by links in a network. Using these correlations, we classify whether a network is assortative or disassortative, indicating the tendency of high-degree nodes to be connected to each other or to low-degree nodes, respectively. We show how to represent this statistical index in a combinatorial expression. We also study other global properties of networks, such as the reciprocity and returnability indices in directed networks.

17.1 Motivation

Characterizing complex networks at a global scale is necessary for many reasons. For example, we can learn about the global topological organization of a given network; it also allows us to compare networks with each other and to obtain information about potential universal mechanisms that give rise to networks with similar structural properties. First we will uncover global topological properties by analysing how frequently the high-degree nodes, or hubs, in a network are connected to each other. We will check the average reciprocity of links in a directed network and the degree to which information departing from a node of a network can return to its source after wandering around the nodes and links. In Chapter 18 we will look at other important global topological properties of networks, too.

17.2 Degree–degree correlation

To measure degree–degree correlation, we record the degrees, k_i and k_j, of the nodes incident to every edge $(i, j) \in E$ in the network. We can then calculate statistics on this set of ordered pairs. If we quantify the linear dependence between k_i and k_j by means of the Pearson correlation coefficient, we will obtain a value $-1 \leq r \leq 1$. The networks where $r > 0$ (positive degree–degree correlation) are known as assortative and those for which $r < 0$ (negative degree–degree correlation) are known as disassortative ones. Those networks for which $r = 0$ are simply know as neutral.

Example 17.1

In Figure 17.1 we have marked each ordered pair of two real-world networks as dots. The social network illustrated is an example of an assortative network, while the mini internet illustrated in the same figure is disassortative.

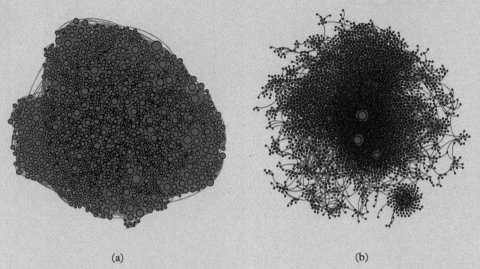

(a)   (b)

Figure 17.1 *(a) Social network of the American corporate elite (b) the internet at the autonomous system level*

Let $e(k_i, k_j)$ be the fraction of links that connect a node of degree k_i to a node of degree k_j. For mathematical convenience, we will consider 'excess degrees', which are simply one less than the degree of the corresponding nodes. Let $p(k_j)$ be the probability that a node selected at random in the network has degree k_j. Then, the Pearson correlation coefficient for the degree–degree correlation is given by

$$r = \frac{1}{\sigma_q^2} \sum_{k_i k_j} k_i\, k_j [e(k_i, k_j) - q(k_i)q(k_j)], \tag{17.1}$$

where

$$q(k_j) = \frac{(k_j + 1)p(k_j + 1)}{\sum_i k_i p(k_i)}, \tag{17.2}$$

represents the distribution of the excess degree of a node at the end of a randomly chosen link and σ_q^2 is the standard deviation of the distribution $q(k_j)$. We call this index the *assortativity coefficient* of a network for obvious reasons. It can be rewritten as

$$r = \frac{\frac{1}{m}\sum_{(i,j)\in E} k_i k_j - \left(\frac{1}{2m}\sum_{(i,j)\in E}(k_i + k_j)\right)^2}{\frac{1}{2m}\sum_{(i,j)\in E}(k_i^2 + k_j^2) - \left(\frac{1}{2m}\sum_{(i,j)\in E}(k_i + k_j)\right)^2}, \tag{17.3}$$

where $m = |E|$.

A revealing property of assortativity can be found by showing that the denominator of (17.3) is nonnegative. We can confirm that

$$\sum_{(i,j)\in E}(k_i^2 + k_j^2) = \sum_i k_i^3 \quad \text{and} \quad \sum_{(i,j)\in E}(k_i + k_j) = \sum_i k_i^2,$$

thus we can write the denominator in (17.3) as

$$\frac{1}{2m}\left[\sum_i k_i^3 - \frac{1}{2m}\left(\sum_i k_i^2\right)^2\right] = \frac{1}{4m^2}\left[\sum_i k_i \sum_i k_i^3 - \left(\sum_i k_i^2\right)^2\right]$$

$$= \frac{1}{8m^2}\left[\sum_{i,j} k_i k_j(k_i^2 + k_j^2) - 2\sum_{i,j}(k_i k_j)^2\right]$$

$$= \frac{1}{8m^2}\left[\sum_{i,j} k_i k_j(k_i - k_j)^2\right] \geq 0.$$

Equality is reached if and only if $k_i = k_j$ for all $i \in V, j \in V$, i.e. for a regular graph. That is, in a regular network the assortativity coefficient is undefined.

In Table 17.1 we give some values of the assortativity coefficient for a few real-world networks.

Table 17.1 *Assortativity coefficients of real-world networks.*

Network	r	Network	r
Drug users	−0.118	Roget thesaurus	0.174
Inter-club friendship	−0.476	Protein structure	0.412
Students dating	−0.119	St. Marks food web	0.118

Example 17.2

The two networks illustrated in Figure 17.2 correspond to food webs. The first represents mostly macroinvertebrates, fishes, and birds associated with an estuarine sea-grass community, *Halodule wrightii*, at St Marks National Wildlife Refuge in Florida. The second represents trophic interactions between birds and predators and arthropod prey of *Anolis* lizards on the island of St Martin in the Lesser Antilles. The first network has 48 nodes and 218 links and the second has 44 nodes and 218 links.

The assortativity coefficient for these two networks are St Marks $r = 0.118$ and St Martin $r = -0.153$. In St Marks, low-degree species prefer to join to other low-degree ones, while high-degree species are preferentially linked to other high-degree ones. On the other hand, in the food web of St Martin, the species with a large number of trophic interactions are preferentially linked to low-degree ones.

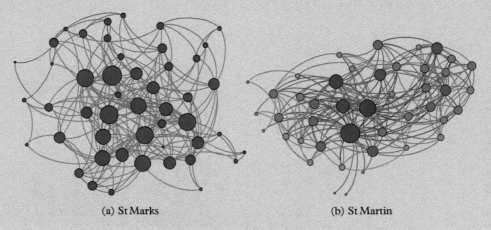

(a) St Marks   (b) St Martin

Figure 17.2 *Illustration of two food webs with different assortativity coefficient values*

17.2.1   Structural interpretation of the assortativity coefficient

Although the idea behind assortativity is very straightforward to describe, there are many cases, even for small networks, in which the measure is hard to rationalize in simple terms.

Example 17.3

Consider the two small networks illustrated in Figure 17.3. The networks are almost identical, except for the fact that the one on the right a path of length 2 instead of one of length 1 attached to the cycle. Despite their close structural similarity, the two have very different degree assortativity.

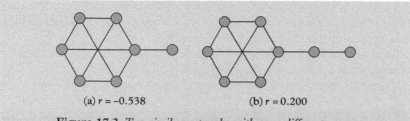

(a) $r = -0.538$ (b) $r = 0.200$

Figure 17.3 *Two similar networks with very different assortativity coefficients*

If we analyse what has been written in the literature about the meaning of assortativity, it is evident that a clear structural interpretation is necessary. For instance, it has been said that in assortatively mixed networks '*the high-degree vertices will tend to stick together in a subnetwork or core group of higher mean degree than the network as a whole*'[1]. However, this may not be visible to the naked eye.

Example 17.4

In the network illustrated in Figure 17.4, the nodes enclosed in the circle indicated with a broken line are among the ones with the highest degree in the network and they are apparently clumped together. Thus we might expect that this network is assortative. However, the assortativity coefficient for this network is $r = -0.304$, showing that it is highly disassortative.

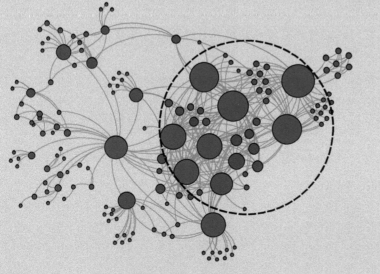

Figure 17.4 *Apparent assortativity within a network*

[1] Newman, M. E. J., Assortative mixing in networks, Physical Review Letters, 89:208701. (2002)

So what can we learn about the structure of complex networks by analysing the assortative coefficient? In other words, what kinds of structural characteristics make some networks assortative or disassortative?

To answer these questions we look for structural interpretations of the different components of (17.3).

The terms common to the numerator and denominator can be rewritten

$$\sum_{(i,j)\in E} (k_i + k_j) = \sum_i k_i^2 = \sum_i (k_i^2 - k_i + k_i) = \sum_i (k_i(k_i - 1)) + \sum_i k_i = 2\,|P_2| + 2m.$$
(17.4)

To interpret the remaining term in the numerator we first rearrange the formula for the number of paths of length 3 to

$$|P_3| = \sum_{(i,j)\in E} (k_i - 1)(k_j - 1) - 3\,|C_3| = \sum_{(i,j)\in E} k_i k_j - \sum_{(i,j)\in E} (k_i + k_j) + m - 3\,|C_3|. \quad (17.5)$$

Then

$$\sum_{(i,j)\in E} k_i k_j = \sum_{(i,j)\in E} (k_i + k_j) - m + 3\,|C_3| + |P_3| = m + 2\,|P_2| + |P_3| + 3\,|C_3|. \quad (17.6)$$

To deal with the final term we first rewrite the expression for the number of star subgraphs $S_{1,3}$ as

$$|S_{1,3}| = \sum_i \binom{k_i}{3} = \frac{1}{6}\sum_i k_i(k_i - 1)(k_i - 2) = \frac{1}{6}\sum_i k_i^3 - \frac{1}{2}\sum_i k_i^2 + \frac{1}{3}\sum_i k_i. \quad (17.7)$$

Then

$$\sum_{(i,j)\in E} (k_i^2 + k_j^2) = \sum_i k_i^3 = 6\,|S_{1,3}| + 3\sum_i k_i^2 - 2\sum_i k_i = 6\,|S_{1,3}| + 6\,|P_2| + 2m.$$
(17.8)

Substituting all these terms into (17.3) gives

$$r = \frac{\dfrac{1}{m}\left(m + 2\,|P_2| + |P_3| + 3\,|C_3|\right) - \dfrac{1}{4m^2}\left(2\,|P_2| + 2m\right)^2}{\dfrac{1}{2m}\left(6\,|S_{1,3}| + 6\,|P_2| + 2m\right) - \dfrac{1}{4m^2}\left(2\,|P_2| + 2m\right)^2} \quad (17.9)$$

which simplifies to

$$r = \frac{|P_3| + 3\,|C_3| - \dfrac{|P_2|^2}{m}}{3\,|S_{1,3}| + |P_2| - \dfrac{|P_2|^2}{m}}. \quad (17.10)$$

Alternatively, let $|P_{r/s}| = |P_r|/|P_s|$ and $|P_1| = m$. Multiply and divide the numerator by $|P_2|$ and we obtain

$$r = \frac{|P_2|\left(|P_{3/2}| + \frac{3|C_3|}{|P_2|} - |P_{2/1}|\right)}{3|S_{1,3}| + |P_2|(1 - |P_{2/1}|)}.$$

(17.11)

Since $C = \dfrac{3|C_3|}{|P_2|}$ is the Newman clustering coefficient,

$$r = \frac{|P_2|\left(|P_{3/2}| + C - |P_{2/1}|\right)}{3|S_{1,3}| + |P_2|\left(1 - |P_{2/1}|\right)}.$$

(17.12)

Both (17.3) and therefore (17.12) have nonnegative denominators so we need only consider the sign of the numerator of (17.12) to determine the nature of assortativity. The conditions for assortative and disassortative mixing can be written as follows.

- A network is assortative ($r > 0$) if and only if $|P_{2/1}| < |P_{3/2}| + C$.
- A network is disassortative ($r < 0$) if and only if $|P_{2/1}| > |P_{3/2}| + C$.

Recall that the clustering coefficient is bounded in the range $0 \le C \le 1$. Examples of assortativity coefficients of real-world networks and the structural parameters involved in the numerator of (17.12) are given in Table 17.2.

As can be seen in Table 17.2, there are some networks for which $|P_{3/2}| > |P_{2/1}|$ and they are assortative independently of the value of their clustering coefficient. In other cases assortativity arises because $|P_{3/2}|$ is slightly smaller than $|P_{2/1}|$ and the clustering coefficient makes $|P_{3/2}| + C > |P_{2/1}|$. It is clear that the role played by the clustering coefficient could be secondary in some cases and determinant in others for the assortativity of a network.

Problem 17.1

Use the combinatorial expression for the assortativity coefficient to show that the path of n nodes of infinite length is neutral, i.e. $r(G) \to 0$ as $n \to \infty$.

In a path $C = 0$ and $|S_{1,3}| = 0$ so the assortativity formula is further simplified to

$$r(P_{n-1}) = \frac{|P_2|\left(|P_{3/2}| - |P_{2/1}|\right)}{|P_2|\left(1 - |P_{2/1}|\right)}.$$

Since $|P_1| = n - 1$, $|P_2| = n - 2$, and $|P_3| = n - 3$,

$$r(P_{n-1}) = \frac{(n-1)(n-3) - (n-2)^2}{(n-1)(n-2) - (n-2)^2} = -\frac{1}{n-2}.$$

So clearly, $\lim_{n\to\infty} r(P_{n-1}) = -\lim_{n\to\infty} \frac{1}{n-2} = 0$.

Table 17.2 *Assortativity coefficient in terms of path ratios and clustering coefficient in complex networks*

Network	$|P_{2/1}|$	$|P_{3/2}|$	C	r
Prison	4.253	4.089	0.288	0.103
Geom	17.416	22.089	0.224	0.168
Corporate	19.42	20.60	0.498	0.268
Roget	9.551	10.081	0.134	0.174
St Marks	10.537	10.464	0.291	0.118
Protein3	4.406	4.45	0.417	0.412
Zachary	6.769	4.49	0.256	−0.476
Drugs	14.576	12.843	0.368	−0.118
Transc Yeast	12.509	3.007	0.016	−0.410
Bridge Brook	22.419	17.31	0.191	−0.664
PIN Yeast	15.66	14.08	0.102	−0.082
Internet	91.00	11.53	0.015	−0.229

Problem 17.2

Use the combinatorial expression for the assortativity coefficient to show that the star of n nodes has the maximum possible disassortativity, i.e. $r(G) = -1$.

In the star with n nodes $C = 0$ and $|P_3| = 0$. Also,

$$|P_1| = n - 1, \quad |P_2| = (n-1)(n-2)/2, \quad |S_{1,3}| = (n-1)(n-2)(n-3)/6.$$

Thus,

$$r(S_{1,n-1}) = \frac{\frac{1}{2}(n-1)(n-2)\left(-\frac{1}{2}(n-2)\right)}{\frac{1}{2}(n-1)(n-2)(n-3) + \frac{1}{2}(n-1)(n-2) - \frac{1}{4}(n-1)(n-2)^2}.$$

This simplifies to

$$r(S_{1,n-1}) = \frac{-\frac{1}{4}(n-1)(n-2)^2}{\frac{1}{2}(n-1)(n-2)\left(\frac{1}{2}(n-2)\right)} = \frac{-\frac{1}{2}(n-2)}{\frac{1}{2}(n-2)} = -1.$$

17.3 Network reciprocity

Another index which is based on the ratio of the number of certain subgraphs in a network is reciprocity. Network reciprocity is defined as

$$r = \frac{L^{\leftrightarrow}}{L}, \tag{17.13}$$

where $L^{\leftrightarrow}$ is the number of links for which a reciprocal exists, and L is the total number of directed links. Thus, if there is a link pointing from A to B, the reciprocity measures the probability that there is also a link pointing from B to A.

A normalized index allows a better characterization of the link reciprocity in a network

$$\rho = \frac{r - \bar{a}}{1 - \bar{a}}, \tag{17.14}$$

where $\bar{a} = L/n(n-1)$. In this case, the number of self-loops are not considered for calculating L, so $L = \sum_{i=1}^{n} \sum_{j=1}^{n} A_{ij} - \sum_{i=1}^{n} A_{ii}$, where A_{ij} is the corresponding entry of the adjacency matrix A of the directed network. It is also clear that $L^{\leftrightarrow} = \text{tr}(A^2)$. According to this index, a directed network can be reciprocal ($\rho > 0$), areciprocal ($\rho = 0$), or antireciprocal ($\rho < 0$). The normalized reciprocity index can be condensed to the formula

$$\rho = \frac{n(n-1)L^{\leftrightarrow} - L^2}{n(n-1)L - L^2}. \tag{17.15}$$

Example 17.5

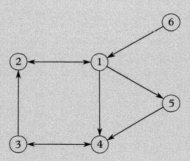

Figure 17.5 *A network with two pairs of reciprocal links*

In the network illustrated in Figure 17.5 there are four reciprocal links ($1-2, 2-1, 3-4$, and $4-3$), thus $L^{\leftrightarrow} = 4$, and the total number of links is $L = 9$. The probability that a link picked randomly in this network is reciprocal is $r = 4/9$. The normalized reciprocity index is then

continued

Example 17.5 *continued*

$$\rho = \frac{30 \cdot 4 - 81}{30 \cdot 9 - 81} = 0.206.$$

In Table 17.3 we illustrate some values of the reciprocity index for real-world networks.

Table 17.3 *Some reciprocal and antireciprocal networks*

Reciprocal	ρ	Areciprocal	ρ	Antireciprocal	ρ
Metab. *E. coli*	0.764	Transc. *E. coli*	−0.0003	Skipwith	−0.280
Thesaurus	0.559	Transc. yeast	$-5.5 \cdot 10^{-4}$	St Martin	−0.130
ODLIS	0.203	Internet	$-5.6 \cdot 10^{-4}$	Chesapeake	−0.057
Neurons	0.158			Little Rock	−0.025

Problem 17.3

A network with n nodes has reciprocity equal to −0.25. How many links should become bidirectional for the network to show reciprocity equal to 0.1?

The reciprocity for this network in its current state is given by

$$\frac{n(n-1)L_1^{\leftrightarrow} - L^2}{n(n-1)L - L^2} = -\frac{1}{4},$$

where $L_1^{\leftrightarrow}$ is the number of bidirectional links in the network. So,

$$L_1^{\leftrightarrow} = \frac{5L^2 - n(n-1)L}{4n(n-1)}.$$

Notice that because $L_1^{\leftrightarrow} > 0$, the number of directed links is bounded by $L > n(n-1)/5$.

Now, let $L_2^{\leftrightarrow}$ be the number of bidirectional links when the reciprocity is 0.1. Then,

$$L_2^{\leftrightarrow} = \frac{9L^2 + n(n-1)L}{10n(n-1)}.$$

Let $\Delta L^{\leftrightarrow} = L_2^{\leftrightarrow} - L_1^{\leftrightarrow}$ be the increase in the number of reciprocal links. Consequently, the number of reciprocal links should be increased by

$$\Delta L^{\leftrightarrow} = \frac{7[n(n-1)L - L^2]}{20n(n-1)}.$$

For instance, if the network has $n = 22, L = 100$, and $L_1^{\leftrightarrow} = 2$, the number of reciprocal links should increase by $\Delta L^{\leftrightarrow} \approx 27$. This means that the new network should have $L_2^{\leftrightarrow} = 29$ in order to have $\rho = 0.1$. You can convince yourself by substituting the values into (17.15).

17.4 Network returnability

Consider a directed network with n nodes, and with adjacency matrix D. The weighted contribution of all returnable cycles in the network gives a global measure of the returnability of information that depart from the nodes. Because no cycles of length smaller than 2 exist we have

$$K_r' = \frac{\text{tr}(D^2)}{2} + \frac{\text{tr}(D^3)}{3} + \cdots + \frac{\text{tr}(D^k)}{k} + \cdots = \text{tr}(\exp(D)) - n - S,$$

where S is the number of self-loops in the network.

The relative returnability can be obtained by normalizing K_r' to

$$K_r = \frac{\text{tr}(\exp(D)) - n - S}{\text{tr}(\exp(A)) - n - S}, \tag{17.16}$$

where A is the adjacency matrix of the same network when all edges are considered to be undirected. This index is bounded in the range $0 \le K_r \le 1$, where the lower bound is obtained for a network with no returnable cycle and the upper bound is obtained for any network with a symmetric adjacency matrix.

Example 17.6

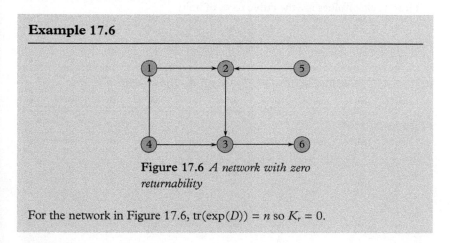

Figure 17.6 *A network with zero returnability*

For the network in Figure 17.6, $\text{tr}(\exp(D)) = n$ so $K_r = 0$.

Problem 17.4

Find the returnability of the directed triads shown in Figure 17.7.

The returnability for these networks can be written

$$K_r = \frac{\text{tr}(\exp(D)) - 3}{\text{tr}(\exp(A)) - 3},$$

where $\text{tr}(\exp(D)) = \sum_{j=1}^{3} \exp(\lambda_j(D))$ and $\text{tr}(\exp(A)) = \exp(2) + 2\exp(-1)$ because the underlying undirected graph is K_3. In order to find the eigenvalues of

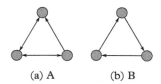

(a) A   (b) B

Figure 17.7 *Two triangles with different returnabilities*

the adjacency matrix of the directed network A, we should find the roots of the polynomial

$$\begin{vmatrix} -\lambda & 0 & 1 \\ 1 & -\lambda & 1 \\ 1 & 1 & -\lambda \end{vmatrix} = -\lambda^3 + 2\lambda + 1 = 0$$

which are $\lambda_1 = \dfrac{\sqrt{5}+1}{2}$, $\lambda_2 = -1$, and $\lambda_3 = \dfrac{\sqrt{5}-1}{2}$. Notice that $\lambda_1 = \varphi$, the golden ratio, and $\lambda_3 = 1 - \varphi$. Hence,

$$K_r(A) = \frac{e^{\varphi} + e^{-1} + e^{1-\varphi} - 3}{e^2 - 2e^{-1} - 3} = 0.567.$$

For the network B we have

$$\begin{vmatrix} -\lambda & 1 & 0 \\ 0 & -\lambda & 1 \\ 1 & 0 & -\lambda \end{vmatrix} = -\lambda^3 + 1 = 0.$$

Thus its eigenvalues are the cubic roots of unity

$$\lambda_j = \cos\left(\frac{2\pi j}{3}\right) + i\sin\left(\frac{2\pi j}{3}\right)$$

for $j = 0, 1, 2$. Substituting into (17.15) we get $K_r(B) = 0.098$.

Example 17.7

In Figure 17.8 we illustrate the returnability of all directed triads.

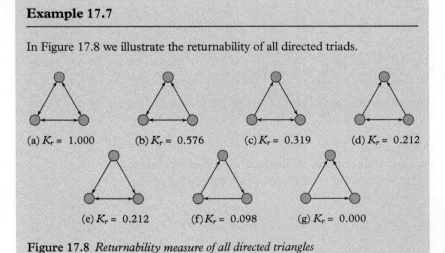

(a) $K_r = 1.000$ (b) $K_r = 0.576$ (c) $K_r = 0.319$ (d) $K_r = 0.212$

(e) $K_r = 0.212$ (f) $K_r = 0.098$ (g) $K_r = 0.000$

Figure 17.8 *Returnability measure of all directed triangles*

....

FURTHER READING

Estrada, E., *The Structure of Complex Networks: Theory and Applications*, Oxford University Press, 2011, Chapters 2.3, 4.5.2, and 5.4.

Newman, M.E.J., *Networks: An Introduction*, Oxford University Press, 2010, Chapter 7.

Newman, M.E.J., 'Assortative Mixing in Networks', *Physical Review Letters*, 89:208701, (2002).

18 Global Properties of Networks II

In this chapter

We introduce the concept of network expansion and the spectral scaling method. These allow us to determine the global topological structure of a complex network. We also study how close to bipartite a network structure is. That is, we characterize how much resemblance a network has to a bipartite graph, which allows us to study real-world situations in which the global bipartivity is somehow 'frustrated'.

18.1 Motivation

Consider two networks with structures such as those displayed in Figure 18.1. Network A displays a very regular, homogeneous type of structure, while network B has a few regions which are more densely connected than others, and which we term structural heterogeneities. Now suppose we use a signal θ that propagates locally through the links of the network from a randomly selected node to other nodes relatively close by. At the same time we emit another signal ϑ, which starts at the same node and propagates on a longer length scale. After a certain time, both signals return to the original node. Because of the homogeneity of network A at both scales (close neighbourhood of a node and global network) the times taken by θ and ϑ to return to the original node are linearly correlated. This is true for any node of the network. However, in B the signal's path will be influenced by the heterogeneity in the network and as a consequence, a lack of correlation between the return times of the two signals is observed. The level of correlation between the two signals characterizes the type of structure that a network has at a global scale. In this chapter, we are going to find a mathematical way to obtain such kinds of correlations for general networks.

18.2 Network expansion properties

The first step towards our characterization of the global topological structure of a network is to find an appropriate definition of network homogeneity. First take

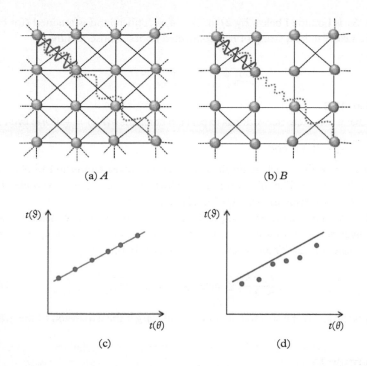

(a) *A*

(b) *B*

(c)

(d)

Figure 18.1 *A homogeneous and a heterogeneous network*

a network $G = (V, E)$ and select a subset $S \subset V$. We designate by ∂S the set of edges that have one endpoint in S and the other outside it, i.e. in $\overline{S}$. The set ∂S is called the boundary of S. Let us now define the following index, known as the *expansion constant* to be

$$\phi(G) = \inf \left\{ \frac{|\partial S|}{|S|}, S \subset V, 0 < |S| \leq \frac{|V|}{2} \right\}. \tag{18.1}$$

If every time that we select a set S the ratio $|\partial S|/|S|$ is about the same, we can conclude that the network is very homogeneous. In graph theory these networks are known as expanders or constant expansion graphs. In contrast, those networks for which the expansion constant changes significantly from one set S to another are characterized by a lack of structural homogeneity, like network B in Figure 18.1.

Problem 18.1

Show that the expansion of a cycle network tends to zero as the size of the network tends to infinity.

The cycle C_n is an example of a connected graph which is divided into two connected components by removing two edges. Note that the boundary of any non-empty set S of nodes from C_n must contain at least two edges and since

$$|S| \leq \frac{|V|}{2} \leq \frac{n}{2},$$

$|\partial S|/|S|$ is bounded below by $2/(n/2) = 4/n$. This bound is attained (for even n) if we take S to be a string of $n/2$ connected nodes. Consequently,

$$\lim_{n\to\infty} \phi(C_n) = \lim_{n\to\infty} \frac{4}{n} = 0.$$

Problem 18.2

Show that a complete network has expansion constant $\phi(K_n) = \dfrac{n}{2}$ if n is even and $\phi(K_n) = \dfrac{n+1}{2}$ if n is odd.

If $|S| = m$ then each node in S has $(n-m)$ edges connected to $\overline{S}$ and so $|\partial S|/|S| = n-m$. The infimum is attained when m is as large as possible. That is $m = n/2$ if n is even and $(n-1)/2$ if n is odd.

A key result in the theory of expander graphs connects the expansion constant to the eigenvalues of the adjacency matrix of the network. Let $\lambda_1 > \lambda_2 \geq \cdots \geq \lambda_n$ be these eigenvalues. Then the expansion factor is bounded by

$$\frac{\lambda_1 - \lambda_2}{2} \leq \phi(G) \leq \sqrt{2\lambda_1(\lambda_1 - \lambda_2)}. \tag{18.2}$$

Thus the larger the spectral gap, $\lambda_1 - \lambda_2$, the larger the expansion of the graph.

Example 18.1

The Petersen graph, illustrated in Figure 18.2, has spectrum

$$\left\{ [3]^1, [1]^5, [-2]^4 \right\}.$$

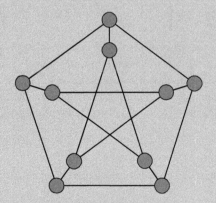

Figure 18.2 *The Petersen graph*

Consequently, $1 \leq \phi(G) \leq 2\sqrt{3}$. In fact, the expansion constant of the Petersen graph is known to be $\phi(G) = 1$.

18.3   Spectral scaling method

Our aim here is to find mathematical analogues of the signals θ and ϑ used in our motivational example. A good candidate for the local signal θ is subgraph centrality. It counts the number of closed walks emanating from a node and penalizes the longer walks more heavily than the shorter ones. It represents a signal that propagates only to the close neighbourhood of a node before returning to it.

Here and elsewhere, we use the odd subgraph centrality instead of $EE(i)$. The reason is simply that $E^{odd}(i)$ accounts for subgraphs which always contain a cycle. It avoids the counting of trivial closed walks which go back and forth without describing a cycle. However, it should be observed that using $EE(i)$ instead of $EE^{odd}(i)$ does not alter the results obtained.

Assuming that the network is neither regular nor bipartite write

$$EE^{odd}(i) = \mathbf{q}_1^2(i)\sinh(\lambda_1) + \sum_{j=2}^{n} \mathbf{q}_j^2(i)\sinh(\lambda_j). \tag{18.3}$$

We now analyse what happens if the network displays structural homogeneity. In this case $\lambda_1 \gg \lambda_2$ and then we can use the approximation

$$EE^{odd}(i) \approx \mathbf{q}_1^2(i)\sinh(\lambda_1). \tag{18.4}$$

Obviously, $\mathbf{q}_1(i)$ is the eigenvector centrality of the node i in the network. In Chapter 15 we established that if a network is not bipartite, the eigenvector centrality is proportional to the number of walks of infinite length that start at node i, $\lim_{k\to\infty} N_k(i)$. Thus (18.4) indicates a power-law relationship between a local characterization of the neighbourhood of the node i (namely $EE^{odd}(i)$) and its global environment ($\mathbf{q}_1(i)$). This is exactly what we are looking for! A power law indicates a self-similarity in a system and in this case it is indicative of the similarities among the local and global topological environments around a given node. We can express (18.4) as

$$\ln \mathbf{q}_1(i) \approx \frac{1}{2}\ln EE^{odd}(i) - \ln\left(\sqrt{\sinh(\lambda_1)}\right). \tag{18.5}$$

For any network, the straight line (18.5) defines the ideal situation in which the local and global environments of all the nodes are highly correlated, i.e. an ideal topological homogeneity in the network. However, the values of $\ln \mathbf{q}_1(i)$ and $\ln EE^{odd}(i)$ can deviate from a straight line if the network is not particularly homogeneous. Such deviations can be quantified, for instance, by measuring the deviation of the eigenvector centrality of the given node from the relationship (18.5) using

$$\Delta \ln \mathbf{q}_1(i) = \ln\left[\frac{\mathbf{q}_1^2(i)\sinh(\lambda_1)}{EE^{odd}(i)}\right]^{1/2}. \tag{18.6}$$

Now we can identify the following four general types of correlation between the local and global environments of the nodes in a network. These can in turn be shown to represent four structural classes of network topology.

18.3.1   Class I: Homogeneous networks

In this case, $\Delta \ln \mathbf{q}_1(i) \approx 0$ for all $i \in V$ so

$$\ln[\mathbf{q}_1^2(i) \sinh(\lambda_1)] \approx \ln EE^{odd}(i), \quad \forall i \in V.$$

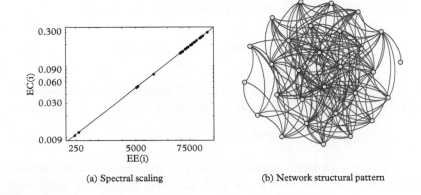

(a) Spectral scaling   (b) Network structural pattern

Figure 18.3 *Spectral scaling for networks in Class I*

18.3.2   Class II: Networks with holes

In this case, $\Delta \ln \mathbf{q}_1(i) < 0$ for all $i \in V$ so

$$\ln[\mathbf{q}_1^2(i) \sinh(\lambda_1)] < \ln EE^{odd}(i), \quad \forall i \in V.$$

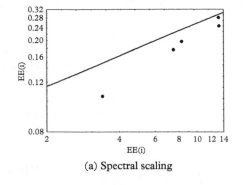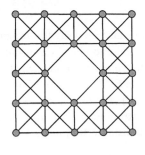

(a) Spectral scaling   (b) Network structural pattern

Figure 18.4 *Spectral scaling for networks in Class II*

18.3.3   Class III: Core–periphery networks

In this case, $\Delta \ln \mathbf{q}_1(i) > 0$ for all $i \in V$ so

$$\ln[\mathbf{q}_1^2(i)\sinh(\lambda_1)] > \ln EE^{odd}(i), \quad \forall i \in V.$$

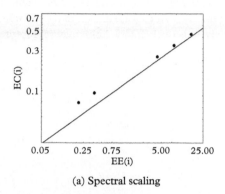

(a) Spectral scaling

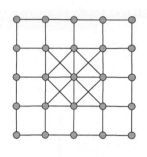

(b) Network structural pattern

Figure 18.5 *Spectral scaling for networks in Class III*

18.3.4   Class IV: Networks with mixed topologies

In this case, $\Delta \ln \mathbf{q}_1(i) < 0$ for some $i \in V$ but $\Delta \ln \mathbf{q}_1(i) > 0$ for other values of i.

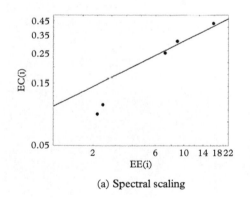

(a) Spectral scaling

(b) Network structural pattern

Figure 18.6 *Spectral scaling for networks in Class IV*

Examples 18.2

(i) Figure 18.7 is a pictorial representation of the food web of St Martin Island and its spectral scaling. The network clearly belongs to Class I.

(ii) Figure 18.8 is an illustration of a protein (left) and its protein residue network (right). In the network, nodes represent the amino acids and the links represent pairs of amino acids interacting physically. The spectral scaling plot in Figure 18.9 shows a clear Class II type for the topology of this network. Proteins are known to fold in 3D, leaving some holes, which in general represent physical cavities where ligands dock. These holes may represent binding sites for potential drugs.

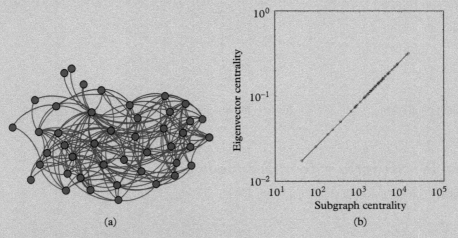

(a)   (b)

Figure 18.7 *Spectral scaling of the St Martin food web. Notice that the scaling perfectly corresponds to a Class I network*

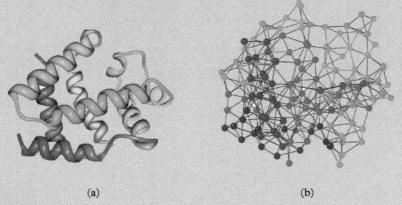

(a)   (b)

Figure 18.8 *Cartoon representation of a protein (a) and its residue interaction network (b)*

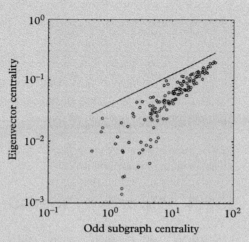

Figure 18.9 *Spectral scaling of the residue network in Figure 18.8(b)*

Problem 18.3

Show that an Erdös–Rényi random network belongs to the first class of homogeneous networks as $n \to \infty$.

Recall that as $n \to \infty$, the eigenvalues of an Erdös–Rényi network satisfy

$$\lambda_1 \to np, \quad \lambda_{j \geq 2} \to 0.$$

Thus, as $n \to \infty$

$$EE^{odd}(i) = \mathbf{q}_1^2(i) \sinh(np) + \sum_{j=2}^{n} \mathbf{q}_j^2(i) \sinh(0) = \mathbf{q}_1^2(i) \sinh(np),$$

which means that

$$\ln \mathbf{q}_1(i) = \frac{1}{2} \ln EE^{odd}(i) - \ln \sqrt{\sinh(np)}$$

and $\Delta \ln \mathbf{q}_1(i) = 0$ as required for a network of Class I.

Problem 18.4

It is known that the largest eigenvalue of the adjacency matrix of a certain protein–protein interaction (PPI) network is 3.8705. Further analysis has shown that the values of the eigenvector and subgraph centrality of the network are 0.2815 and 0.5603, respectively. Is this PPI network an example of a homogeneous network? Or does it belong to the class of networks with holes?

If the PPI network were in Class I, then $\Delta \ln \mathbf{q}_i(i) \approx 0$ for all $i \in V$. Using the given data, we have

$$\Delta \ln \mathbf{q}_1(i) = \ln \left[\frac{0.2815^2 \sinh(3.8705)}{0.5603} \right]^{1/2} = 0.6105,$$

which is far from zero. Thus the network does not belong to the class of homogeneous networks. In addition, because $\Delta \ln \mathbf{q}_1(i) > 0$ for at least one node, the network cannot belong to the class of networks with holes. The only possibilities that remain are that the network is in Class III or Class IV but without additional data we cannot determine which.

18.4   Bipartivity measures

Bipartivity arises naturally in many real-world networks. Think for instance of a network with nodes $V = V_1 \bigcup V_2$ in which V_1 represents a set of buyers and V_2 represents a set of sellers. In general, there are only links connecting buyers to sellers. However, in some cases there will be sellers who make purchases from other sellers. In this case, the network is no longer bipartite but it could be possible that it still has some resemblances to a bipartite network. In this section our aim is to characterize the global structure of a network in such a way as to introduce a grey scale in which at the two extremes there are the networks which are bipartite or extremely nonbipartite, and in the middle all the networks that display some degree of bipartity.

The simplest and more intuitive way of defining the bipartivity of a network is to account for the fraction of links that 'destroy' the bipartity in the network. That is, to calculate the minimum number of links we have to remove to make the network perfectly bipartite. Let m_d be the number of such links in the network and let m be the total number of links. Then, the bipartivity of a network can be measured using

$$b_c = 1 - \frac{m_d}{m}, \tag{18.7}$$

where the subindex c is introduced to indicate that this index may have to be found computationally. That is, we need to search for links that destroy the bipartity in the network in a computationally intensive way. We aim to find the best partition of the nodes into two almost disjoint subsets. Once we have found such a 'best bipartition', we can simply count the number of links that are connecting nodes in the same set. Although the method is very simple and intuitive, to state, the computation of this index is an example of an NP-complete problem.[1] Computational approximations for finding this index have been reported in the literature but we are not going to explain them here. Instead we will consider other approaches that exploit some of the spectral properties of bipartite networks.

[1] In short, this means that it is extremely costly to compute for even fairly modestly sized networks.

Recall that a bipartite network does not contain any odd-length cycle. Because every closed walk of odd length involves at least one cycle of odd length, we also know that a bipartite network does not contain any closed walks of odd length. Consequently, we can identify bipartivity if $\text{tr}(\sinh(A)) = 0$. Or, because

$$\text{tr}(e^A) = \text{tr}(\sinh(A)) + \text{tr}(\cosh(A)),\tag{18.8}$$

we have bipartivity if and only if

$$\text{tr}(e^A) = \text{tr}(\cosh(A)).\tag{18.9}$$

Using these simple facts we can design an index to account for the degree of bipartivity of a network by taking the proportion of even closed walks to the total number of closed walks in the network giving

$$b_s = \frac{\text{tr}(\cosh(A))}{\text{tr}(e^A)} = \frac{\displaystyle\sum_{j=1}^{n} \cosh(\lambda_j)}{\displaystyle\sum_{j=1}^{n} \exp(\lambda_j)}.\tag{18.10}$$

It is evident that $b_s \leq 1$ with equality if and only if the network is bipartite.

Example 18.3

We consider the effect of adding a new edge, e, to a network G with spectral bipartivity index $b_s(G)$. Let us add a new edge to G and calculate $b_s(G + e)$.

Since $\cosh x > \sinh x$ for all x,

$$EE^{even} = \sum_{j=1}^{n} \cosh(\lambda_j) > \sum_{j=1}^{n} \sinh(\lambda_j) = EE^{odd}.\tag{18.11}$$

Let a and b be the contributions of e to the even and odd closed walks in $G + e$, respectively and assume that $b \geq \alpha a$, where $\alpha = \alpha(G)$ is the ratio $\left(\sum_{j=1}^{n} \sinh(\lambda_j)\right) / \left(\sum_{j=1}^{n} \cosh(\lambda_j)\right)$. Then $bEE^{even} \geq aEE^{odd}$. See the appendix for more detail.

Adding $EE^{even}(a + EE)$ to each side gives

$$EE^{even}(a + b + EE) \geq EE(EE^{even} + a),$$

which can be written as

$$b_s(G) = \frac{EE^{even}}{EE} \geq \frac{EE^{even} + a}{EE + a + b} = b_s(G + e).\tag{18.12}$$

This means that adding a link that makes a larger contribution to odd closed walks than even ones will always decrease the value of b_s with respect to that of the original network.

Suppose we start from a bipartite network with $b_s = 1$ and add new links. How small can we make b_s? And for what network?

Problem 18.5

Show that $b_s \to 1/2$ in K_n as $n \to \infty$.

The eigenvalues of K_n are $n - 1$ with multiplicity one and -1 with multiplicity $n - 1$ so

$$b_s(K_n) = \frac{\cosh(n-1) + (n-1)\cosh(-1)}{\exp(n-1) + (n-1)\exp(-1)} \to \frac{1}{2}$$

as $n \to \infty$.

Of all graphs with n nodes, the complete graph is the one with the largest number of odd cycles. So $\lim_{n\to\infty} b_s(K_n) = 1/2$ is the minimum value of b_s that any network can take.

Another way of accounting for the global bipartivity of a network is to consider the difference of the number of closed walks of even and odd length, and then to normalize the index by the sum of closed walks. That is,

$$b_e = \frac{\sum_{j=1}^{n} \cosh(\lambda_j) - \sum_{j=1}^{n} \sinh(\lambda_j)}{\sum_{j=1}^{n} \cosh(\lambda_j) + \sum_{j=1}^{n} \sinh(\lambda_j)} = \frac{\mathrm{tr}(\exp(-A))}{\mathrm{tr}(\exp(A))} = \frac{\sum_{j=1}^{n} \exp(-\lambda_j)}{\sum_{j=1}^{n} \exp(\lambda_j)}. \qquad (18.13)$$

Clearly, $0 \le b_e \le 1$.

Problem 18.6

Show that $b_e = 1$ for any bipartite network and that $b_e(K_n) \to 0$ as $n \to \infty$.

For a bipartite network the spectrum of the adjacency matrix is symmetrically distributed around zero. Since sinh is an odd function.

$$\sum_{j=1}^{n} \sinh(\lambda_j) = 0$$

and so $b_e = 1$.

Using the eigenvalues of the adjacency matrix of a complete graph we have

$$b_e(K_n) = \frac{\exp(1-n) + (n-1)\exp(1)}{\exp(n-1) + (n-1)\exp(-1)} \to 0, \qquad (18.14)$$

as $n \to \infty$.

Problem 18.7

Suppose that we repeatedly add edges to a network G and each time we do we add more odd walks than even walks. Show that b_e decreases monotonically with the addition of edges.

Suppose again that a and b are the contributions of the new edge e to even and odd closed walks, respectively, and $b \geq \alpha a$. Then

$$b_e(G + e) = \frac{(EE^{even} + a) - (EE^{odd} + b)}{EE + a + b} = \frac{EE^{even} + a}{EE + a + b} - \frac{EE^{odd} + b}{EE + a + b}.$$

Adding $(b + EE)EE^{odd}$ to each side of the inequality $bEE^{even} \geq aEE^{odd}$ gives

$$EE(EE^{odd} + b) \geq EE^{odd}(EE + a + b),$$

which means that

$$\frac{EE^{odd} + b}{EE + a + b} \geq \frac{EE^{odd}}{EE}.$$

We have previously shown that

$$\frac{EE^{even}}{EE} \geq \frac{EE^{even} + a}{EE + a + b}.$$

Combining these last two inequalities gives

$$b_e(G) = \left(\frac{EE^{even}}{EE} - \frac{EE^{odd}}{EE} \right) \geq \left(\frac{EE^{even} + a}{EE + a + b} - \frac{EE^{odd} + b}{EE + a + b} \right) = b_e(G + e),$$

as desired.

Examples 18.4

(i) In Figure 18.10 we show how the values of the two spectral indices of bipartivity change as we add links to the complete bipartite network $K_{2,3}$.

(ii) Figure 18.11a illustrates a food web among invertebrates in an area of grassland in England. The values of the spectral bipartivity indices for this network are $b_s = 0.766$ and $b_e = 0.532$, both indicating that the network displays some bipartite-like structure due to the different trophic layers of interaction among the species. We will see how to find such bipartitions in Chapter 21.

(iii) Figure 18.11b illustrates an electronic sequential logic circuit where nodes represent logic gates. The values of the spectral bipartivity indices for this network are $b_s = 0.948$ and $b_e = 0.897$, both indicating that the network displays a high bipartivity.

continued

Examples 18.4 *continued*

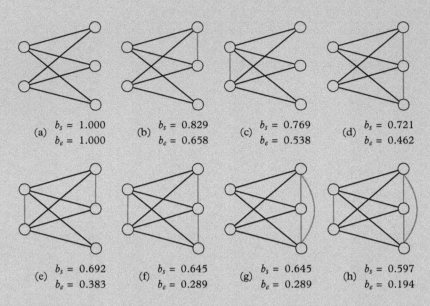

Figure 18.10 *Monotonic decay of bipartivity indices as edges are added*

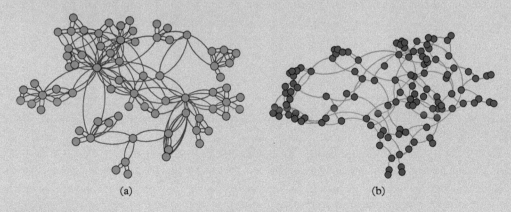

Figure 18.11 *(a) A food network in English grassland (b) An electronic circuit network*

Problem 18.8

Find the conditions for which an Erdös–Rényi (ER) network will have $b_e \approx 1$ and those for which $b_e \approx 0$ when the number of nodes, n, is sufficiently large.

As $n \to \infty$ the eigenvalues of an ER network are given by $\lambda_1 = np, \lambda_{j \geq 2} = 0$. Thus,

$$b_e(ER) = \frac{n - 1 + e^{-np}}{n - 1 + e^{np}} = \frac{n - 1}{n - 1 + e^{np}} + \frac{e^{-np}}{n - 1 + e^{np}} .$$

There is a regime for which the first term on the right-hand side is dominant in the limit and another in which it is the second. In the first case the limit approaches 1 if $\exp(np) \to 0$. In the second case, the limit will approach 0 if $\exp(np) \to \infty$. Consequently, $b_e(ER) \approx 1$ if $\exp(np) \ll n - 1$. This condition is equivalent to

$$p \ll \frac{\ln(n - 1)}{n} . \tag{18.15}$$

Sometimes $p < \dfrac{\ln(n - 1)}{n}$ is sufficient. For example, in an ER network with 1,000 nodes and 2,990 links we measure $b_e(ER) \approx 0.9$. Notice that here $np \approx 6$ while $\ln(n - 1) \approx 7$. However in most cases where (18.15) is not satisfied we can expect very low bipartivity in ER networks.

....

FURTHER READING

Estrada, E., Spectral scaling and good expansion properties in complex networks, Europhys. Lett. **73**:649–655, 2006.

Hoory, S., Linial, N. and Wigderson, A., Expander graphs and their applications. Bulletin of the American Mathematical Society **43**:439–561, 2006.

Sarnak, P., What is an Expander? Notices of the AMS, **51**:761–763, 2004.

19 Communicability in Networks

In this chapter

We introduce the concept of communicability between two nodes in a network. It accounts for all the possible routes (i.e. walks of a given length) that connect the corresponding pair of nodes and gives more weight to the shorter than to the longer ones. We define the communicability distance between two nodes, which accounts for how well connected the nodes are in a network and show applications.

19.1 Motivation

The transmission of information is one of the principal functions of complex networks. Such communication among the nodes of a network can represent the interchange of thoughts or opinions in social networks, the transfer of information from one molecule or cell to another by means of chemical, electrical, or other kind of signal, or the routes of transportation of any material. It is intuitive to think that this communication mainly takes place by using the shortest route connecting a pair of nodes. That is, to assume that information is transmitted through the shortest paths of a network. This is represented in Figure 19.1 (a) by the blue path between the two marked nodes in the network. However, in any network different from a tree, there are many other routes for communication between any pair of nodes. This abundance of alternative routes is of great relevance when there are failures in some of the links in the shortest path, or simply if there is

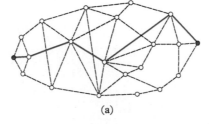
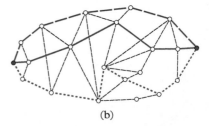

Figure 19.1 *(a) The shortest path between the two coloured nodes (b) Alternative routes when an edge fails*

(a)　　　　　　　　　(b)

heavy traffic along some routes. Some of these alternative routes are illustrated in Figure 19.1 (b) where the central link of the shortest path connecting the two marked nodes has been removed.

We conclude that the nodes in a network have improved communication if there are many relatively short alternative routes between them, other than the shortest path. The larger the number of these alternative routes the better the *communicability* between the corresponding pair of nodes. This is the topic of this chapter.

19.2 Network communicability

In order to account for the routes of communication between two nodes in a network we appeal again to the concept of walk. That is, the communicability of two nodes is characterized by the number of walks starting at one node and ending at the other and the relative importance of the walks connecting both nodes in terms of their length. This allows us to use the following mathematical definition for the communicability of the nodes p and q:

$$G_{pq} = \sum_{k=0}^{\infty} c_k (A^k)_{pq}, \qquad (19.1)$$

where the coefficients c_k must fulfil the same requirements stated when we introduced subgraph centrality. By selecting $c_k = 1/k!$ we obtain

$$G_{pq} = \sum_{k=0}^{\infty} \frac{(A^k)_{pq}}{k!} = (\exp(A))_{pq}. \qquad (19.2)$$

Recall that G_{pp} measures the subgraph centrality of a node. Using the spectral decomposition of the adjacency matrix the communicability function can be expressed as

$$G_{pq} = \sum_{j=1}^{n} \mathbf{q}_j(p)\mathbf{q}_j(p) \exp(\lambda_j). \qquad (19.3)$$

Other communicability functions can also be obtained by choosing the constants c_k in other ways. For example,

$$G_{pq}^{odd} = (\sinh(A))_{pq}, \qquad (19.4)$$
$$G_{pq}^{even} = (\cosh(A))_{pq}, \qquad (19.5)$$
$$G_{pq}^{res} = \left((I - \alpha A)^{-1}\right)_{pq}, \quad 0 < \alpha < 1/\lambda_1. \qquad (19.6)$$

However, hereafter we will refer to (19.2) as the communicability function.

The communicability function must reproduce some of our intuition about the transmission of information in a network. For instance, we expect a very high communicability between nodes in a complete graph. Also we should expect that the communicability between the endpoints of a path decays asymptotically to zero as the length of the path increases. This is easily established. For example, since

$$\sigma(K_n) = \left\{ [n-1]^1, [-1]^{n-1} \right\},$$

we can write

$$G_{pq}(K_n) = \mathbf{q}_1(p)\mathbf{q}_1(q)e^{n-1} + e^{-1} \sum_{j=2}^{n} \mathbf{q}_j(p)\mathbf{q}_j(p). \tag{19.7}$$

And since the eigenvector matrix Q has orthonormal rows and columns with $\mathbf{q}_1 = e/\sqrt{n}$,

$$G_{pq}(K_n) = \frac{e^{n-1}}{n} + e^{-1}\left(1 - \frac{1}{n}\right) \to \infty$$

as $n \to \infty$.

Now, we turn to the path of n edges P_n. In this case, we know that the eigenvalues are

$$\lambda_j = 2\cos\left(\frac{j\pi}{n+1}\right), \tag{19.8}$$

and the entries of the eigenvectors are

$$\mathbf{q}_j(p) = \sqrt{\frac{2}{n+1}}\sin\frac{jp\pi}{n+1}. \tag{19.9}$$

So

$$G_{pq}(P_n) = \sum_{j=1}^{n} \left(\sqrt{\frac{2}{n+1}}\sin\frac{jp\pi}{n+1}\right)\left(\sqrt{\frac{2}{n+1}}\sin\frac{jq\pi}{n+1}\right)e^{2\cos\left(\frac{j\pi}{n+1}\right)}$$

$$= \frac{2}{n+1}\sum_{j=1}^{n}\left(\sin\frac{jp\pi}{n+1}\right)\left(\sin\frac{jq\pi}{n+1}\right)e^{2\cos\left(\frac{j\pi}{n+1}\right)}.$$

Using the standard trigonometric identity $2\sin\theta\sin\vartheta = \cos(\theta-\vartheta) - \cos(\theta+\vartheta)$,

$$G_{pq}(P_n) = \frac{1}{n+1}\sum_{j=1}^{n}\left(\cos\frac{j\pi(p-q)}{n+1} - \cos\frac{j\pi(p+q)}{n+1}\right)e^{2\cos\left(\frac{j\pi}{n+1}\right)}$$

$$= \frac{1}{n+1}\sum_{j=1}^{n}\left\{\cos\left(\frac{j\pi(p-q)}{n+1}\right)e^{2\cos\left(\frac{j\pi}{n+1}\right)} - \frac{1}{n+1}\cos\left(\frac{j\pi(p+q)}{n+1}\right)e^{2\cos\left(\frac{j\pi}{n+1}\right)}\right\}.$$

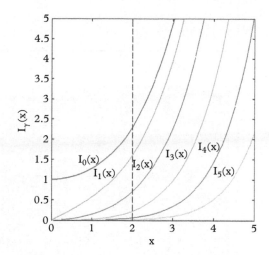

Figure 19.2 *Plot of the modified Bessel functions of the first kind. The vertical line at x = 2 helps illustrate that $I_\gamma(2) \to 0$*

For $j = 1, 2, \ldots, n$ the angles $j\pi/(n+1)$ uniformly cover the interval $[0, \pi]$, justifying the integral approximation

$$G_{pq}(P_n) \approx \frac{1}{\pi} \int_0^\pi \cos(\theta(p-q))e^{2\cos\theta}\, d\theta - \frac{1}{\pi} \int_0^\pi \cos(\theta(p+q))e^{2\cos\theta}\, d\theta, \quad (19.10)$$

where $\theta = j\pi/(n+1)$. From integral tables,

$$\frac{1}{\pi} \int_0^\pi \cos(\gamma\theta)e^{x\cos\theta}\, d\theta = I_\gamma(x), \quad (19.11)$$

which is a modified Bessel function of the first kind. Hence,

$$G_{pq}(P_n) \approx I_{p-q}(2) - I_{p+q}(2). \quad (19.12)$$

Figure 19.2 illustrates Bessel functions of the first kind. We see that as $\gamma \to \infty, I_\gamma(2) \to 0$ very quickly. Consequently, $G_{n1}(P_n) \approx (I_{n-1}(2) - I_{n+1}(2)) \to 0$ as $n \to \infty$.

For instance, $G_{1,5}(P_5) = 0.048$ and $G_{1,10}(P_{10}) = 2.98 \times 10^{-6}$.

Example 19.1

Consider the network of interconnections between the different regions of the visual cortex of the macaque. Figure 19.3 illustrates a planar map of the regions of the visual cortex (a) and the map of communicability between pairs of regions (b). As can be seen, there are a few small regions which communicate intensely.

continued

Example 19.1 *continued*

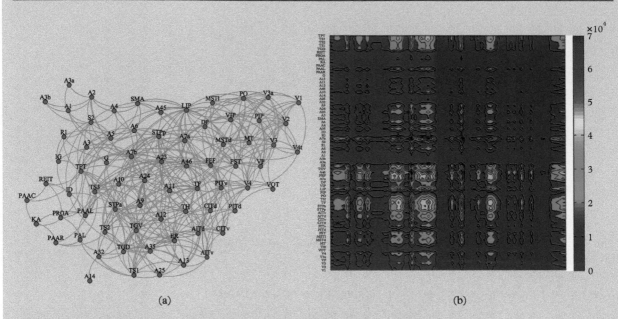

(a)   (b)

Figure 19.3 *(a) A representation of the macaque cortex (b) The communicability between regions*

Problem 19.1

Let $S_{1,n-1}$ be a star with n nodes in which we have labelled the central node as 1. Given that the eigenvectors associated with the largest and the smallest eigenvalues are

$$\mathbf{q}_1 = \frac{1}{\sqrt{2(n-1)}} \left[\sqrt{n-1} \quad 1 \quad \cdots \quad 1 \right]^T$$

and

$$\mathbf{q}_n = \frac{1}{\sqrt{2(n-1)}} \left[-\sqrt{n-1} \quad 1 \quad \cdots \quad 1 \right]^T$$

and that the spectrum is

$$\sigma(S_{1,n-1}) = \left\{ \left[\sqrt{n-1} \right]^1, [0]^{n-2}, \left[-\sqrt{n-1} \right]^1 \right\},$$

find expressions for the communicability between any pair of nodes in the network.

In the star graph there are two nonequivalent types of pairs of nodes. One is formed by the central node and any of the nodes of degree one, the other is formed by any pair of nodes of degree one. Let us designate the communicability between the first type by $G_{p1}(S_{1,n-1})$ and the second by $G_{pq}(S_{1,n-1})$.

By substituting the values of the eigenvalues and eigenvectors into the expression for the communicability we have

$$G_{p1}(S_{1,n-1}) = \frac{1}{\sqrt{2}}\frac{1}{\sqrt{2(n-1)}}e^{\sqrt{n-1}} - \frac{1}{\sqrt{2}}\frac{1}{\sqrt{2(n-1)}}e^{-\sqrt{n-1}} + \sum_{j=2}^{n-1}\mathbf{q}_j(1)\mathbf{q}_j(p), \ p \neq 1. \tag{19.13}$$

From the orthonormality of the eigenvectors we deduce that

$$0 = \sum_{j=1}^{n}\mathbf{q}_j(1)\mathbf{q}_j(p) = \sum_{j=2}^{n-1}\mathbf{q}_j(1)\mathbf{q}_j(p) + \mathbf{q}_1(1)\mathbf{q}_1(p) + \mathbf{q}_n(1)\mathbf{q}_n(p). \tag{19.14}$$

Thus,

$$\sum_{j=2}^{n-1}\mathbf{q}_j(1)\mathbf{q}_j(p) = 0, \tag{19.15}$$

and if $p \neq 1$,

$$G_{p1}(S_{1,n-1}) = \frac{1}{\sqrt{n-1}}\left(\frac{e^{\sqrt{n-1}} - e^{-\sqrt{n-1}}}{2}\right) = \frac{\sinh\left(\sqrt{n-1}\right)}{\sqrt{n-1}}.$$

Similarly,

$$G_{pq}(S_{1,n-1}) = \frac{1}{2(n-1)}e^{\sqrt{n-1}} + \frac{1}{2(n-1)}e^{-\sqrt{n-1}} + \sum_{j=2}^{n-1}\mathbf{q}_j(p)\mathbf{q}_j(q). \tag{19.16}$$

Again, from orthonormality considerations, if p and q are different and not equal to 1,

$$\sum_{j=2}^{n-1}\mathbf{q}_j(p)\mathbf{q}_j(q) = -\frac{1}{n-1}. \tag{19.17}$$

Thus for $p, q > 1, p \neq q$,

$$G_{pq}(S_{1,n-1}) = \frac{1}{n-1}\left[\cosh(\sqrt{n-1}) - 1\right]. \tag{19.18}$$

Notice that the communicability between the central node and any other node is determined only by walks of odd length while that between any two noncentral nodes is determined by even length walks only.

19.3 Communicability distance

We start by recalling the results obtained by Milgram in his 1967 experiment that gave rise to the concept of 'small-world' networks. In that experiment it was clearly observed that a letter which arrives at its destination travels through a very small number of steps. However, letters can be lost due to the fact that nodes along the route may be involved in many transitive relations.

The first of these observations describes a situation when a letter travels along (close to) shortest paths connecting the origin and the destination, if we assume that the average path length in the network is very small. The second observation is accounted for by the fact that the nodes in the network display a high clustering coefficient. As we have seen before, these are the two main ingredients of the Watts–Strogatz model of 'small-world' networks.

We can devise a simple strategy in order to optimize the chances that a letter arrives at its destination in a given network. The sender is supposed to know nothing about the global structure of the network. She only knows who her acquaintances are in that network. Instead of selecting one of these acquaintances at random, the sender is supposed to give her letter to one of the most connected of her ties. If the new receiver of the letter also uses the same strategy the result is that the letter travels through highly connected nodes with high probability. Highly connected nodes have a large number of paths crossing through them. Thus, the probability that the shortest path connecting the origin and the destination of the letter passes through such highly connected nodes is very high (see Figure 19.4(a)). Consequently, the letter should arrive in a small number of steps if it is sent via the highly connected acquaintances of each node. However, it is also expected that those highly connected nodes are involved in a large number of triangles due to the transitivity of the relations. Consequently, sending the letter through these nodes also increases the chances that the letter is lost (see Figure 19.4(b)). In closing, sending the letter through the highly connected acquaintances of a node both increases the chance that the letter arrives at the destination because it uses the shortest path from the origin to the destination; and increases the chance that the letter is lost due to the high transitivity in which those nodes are involved.

This apparent paradox in the communication between two nodes when the shortest path route is used motivates an index that accounts for communication

Figure 19.4 *(a) Shortest path between source and target passes through the hub (b) Walks starting and ending at the hub*

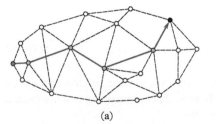

(a)

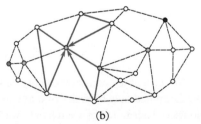

(b)

routes that maximize the communication between a pair of nodes but reduce the disruption in the communication due to transitivity of the relationships.

In order to identify routes that maximise the communication between nodes p and q, it is intuitive to think about the communicability function G_{pq}, which accounts for the amount of information that departs from p and successfully arrives at q. The 'disruption' in the communication is represented by the information that departs from p and after wandering around the nodes of the network returns again to p. A natural index to account for this disrupted information is G_{pp}. If we assume that the communication between p and q is bidirectional, there is also disruption in the information sent from q, accounted for by G_{qq}.

We define an index that accounts for the amount of information disrupted minus the amount of information that arrives at its destination with

$$\xi'_{pq} \overset{\text{def}}{=} G_{pp} + G_{qq} - 2G_{pq}. \tag{19.19}$$

Minimizing ξ'_{pq} represents the case in which we minimize loss of information and at the same time maximize the information that finally arrives at its destination. We can show that ξ'_{pq} can be viewed as the square of an appropriately defined Euclidean distance between the nodes p and q of the network.

Let $A = QDQ^T$ be the spectral decomposition of the adjacency matrix of a network and denote the pth row of Q by $\mathbf{u}_p^T$. Then we can write

$$\xi'_{pq} = (\mathbf{u}_p - \mathbf{u}_q)^T e^D (\mathbf{u}_p - \mathbf{u}_q), = \left\{ (e^{D/2}(\mathbf{u}_p - \mathbf{u}_q)) \right\}^T \left\{ (e^{D/2}(\mathbf{u}_p - \mathbf{u}_q)) \right\}$$

since $e^D = e^{D/2}e^{D/2}$. Writing $\mathbf{x}_p = e^{D/2}\mathbf{u}_p$,

$$\xi'_{pq} = (\mathbf{x}_p - \mathbf{x}_q)^T (\mathbf{x}_p - \mathbf{x}_q) = \|\mathbf{x}_p - \mathbf{x}_q\|^2,$$

hence

$$\xi_{pq} = \sqrt{\xi'_{pq}} = \|\mathbf{x}_p - \mathbf{x}_q\| \tag{19.20}$$

is a Euclidean distance. For obvious reasons, we will call ξ_{pq} the *communicability distance* between the nodes p and q of a graph.

We can define an analogue of the distance matrix for the communicability distance as follows. Let $\mathbf{s} = \begin{bmatrix} EE_{11} & EE_{22} & \cdots & EE_{nn} \end{bmatrix}^T$ be a column vector of the subgraph centralities of every node in the graph and let

$$M = \mathbf{s}\mathbf{e}^T + \mathbf{e}\mathbf{s}^T - 2\exp(A). \tag{19.21}$$

Then the *communicability distance matrix* of a network is given by

$$X(G) = \sqrt[\circ]{M}, \tag{19.22}$$

where $\sqrt[\circ]{}$ is the componentwise square root.

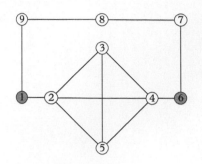

Figure 19.5 *Calculate the communicability of the coloured nodes*

Problem 19.2

Consider the network illustrated in Figure 19.5. Find the routes connecting nodes 1 and 6 with shortest path and communicability distances.

First, we identify the paths connecting 1 and 6. There are 6, namely,

$$1 \to 2 \to 4 \to 6 \quad 1 \to 2 \to 5 \to 4 \to 6$$
$$1 \to 2 \to 3 \to 5 \to 4 \to 6 \quad 1 \to 2 \to 5 \to 3 \to 4 \to 6$$
$$1 \to 2 \to 3 \to 4 \to 6 \quad 1 \to 9 \to 8 \to 7 \to 6.$$

In order to find the one with the shortest path distance, we simply sum the shortest path distances between the pairs of nodes forming the path (in this unweighted network the distance between adjacent nodes is 1). It is obvious that the shortest path distance is 3 for $1 \to 2 \to 4 \to 6$.

For the communicability distance we proceed in a similar way and calculate $\xi_{12} + \xi_{24} + \xi_{46}, \xi_{12} + \xi_{23} + \xi_{34} + \xi_{46}$, and so on. In order to calculate the communicability distance we need

$$G = e^A = \begin{bmatrix} 2.482 & 2.597 & 1.625 & 1.704 & 1.625 & 0.465 & 0.299 & 0.709 & 1.628 \\ 2.597 & 6.530 & 5.510 & 5.842 & 5.510 & 1.704 & 0.470 & 0.333 & 0.905 \\ 1.625 & 5.510 & 5.558 & 5.510 & 5.190 & 1.625 & 0.416 & 0.167 & 0.416 \\ 1.704 & 5.842 & 5.510 & 6.530 & 5.510 & 2.597 & 0.905 & 0.333 & 0.470 \\ 1.625 & 5.510 & 5.190 & 5.510 & 5.558 & 1.625 & 0.416 & 0.167 & 0.416 \\ 0.465 & 1.704 & 1.625 & 2.597 & 1.625 & 2.482 & 1.628 & 0.709 & 0.299 \\ 0.299 & 0.470 & 0.416 & 0.905 & 0.416 & 1.628 & 2.285 & 1.594 & 0.704 \\ 0.709 & 0.333 & 0.167 & 0.333 & 0.167 & 0.709 & 1.594 & 2.280 & 1.594 \\ 1.628 & 0.905 & 0.416 & 0.470 & 0.416 & 0.299 & 0.704 & 1.594 & 2.285 \end{bmatrix}.$$

Note that **s** is the diagonal of G and from (19.21) and (19.22) we obtain

$$X(G) = \begin{bmatrix} 0.000 & 1.954 & 2.188 & 2.367 & 2.188 & 2.008 & 2.042 & 1.829 & 1.230 \\ 1.954 & 0.000 & 1.033 & 1.173 & 1.033 & 2.367 & 2.806 & 2.854 & 2.647 \\ 2.188 & 1.033 & 0.000 & 1.033 & 0.858 & 2.188 & 2.648 & 2.739 & 2.648 \\ 2.367 & 1.173 & 1.033 & 0.000 & 1.033 & 1.954 & 2.647 & 2.854 & 2.806 \\ 2.188 & 1.033 & 0.858 & 1.033 & 0.000 & 2.188 & 2.648 & 2.739 & 2.648 \\ 2.008 & 2.367 & 2.188 & 1.954 & 2.188 & 0.000 & 1.230 & 1.829 & 2.042 \\ 2.042 & 2.806 & 2.648 & 2.647 & 2.648 & 1.230 & 0.000 & 1.174 & 1.778 \\ 1.829 & 2.854 & 2.739 & 2.854 & 2.739 & 1.829 & 1.174 & 0.000 & 1.174 \\ 1.230 & 2.647 & 2.648 & 2.806 & 2.648 & 2.042 & 1.778 & 1.174 & 0.000 \end{bmatrix}.$$

In this case, $\xi_{12} + \xi_{24} + \xi_{46} = 5.081$ but $\xi_{19} + \xi_{98} + \xi_{87} + \xi_{76} = 4.808$. This means that according to the communicability distance, the route $1 \to 9 \to 8 \to 7 \to 6$ is shorter than $1 \to 2 \to 4 \to 6$ and it can be confirmed that no other route has shorter communicability distance. The two routes are marked in Figure 19.6.

In order to understand the differences between the two routes we just need to recall the definition of communicability distance $\xi'_{pq} = G_{pp} + G_{qq} - 2G_{pq}$.

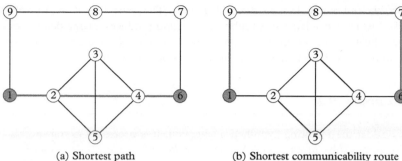

(a) Shortest path	(b) Shortest communicability route

Figure 19.6 *A contrast between distance and communicability minimization*

It indicates a route that maximizes the communicability between the two nodes and minimizes the disruption. The route $1 \rightarrow 9 \rightarrow 8 \rightarrow 7 \rightarrow 6$ certainly reduces the chances of getting lost along the way without increasing the length of the route excessively.

A good analogy for understanding the differences between the shortest path and the communicability distance is the following. Suppose every node is represented by a ball of mass proportional to its subgraph centrality. Then a node participating in many small subgraphs will have a very large mass. Now place the network onto a rubber sheet which will be deformed according to the masses of the corresponding nodes. Then to travel from one node to another we need to follow the 'geodesic' paths, which include the deformations of the sheet. Consequently, as illustrated in Figure 19.7, going from one node to another by using a route that involves nodes of large masses (large subgraph centrality) will increase the length of the trajectory greatly in comparison with those routes involving low mass nodes only. In the picture, there are two alternative routes between nodes 1 and 3. The route $1 \rightarrow 2 \rightarrow 3$ involves node 2 which has very low subgraph

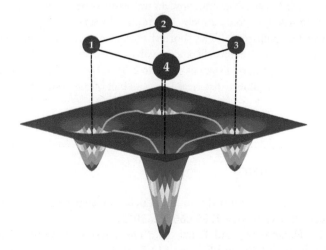

Figure 19.7 *Representation of communicability as the level of deformation of a surface*

centrality and barely deforms the rubber sheet. The route $1 \rightarrow 4 \rightarrow 3$ involves node 4 which has a large subgraph centrality and produces a large deformation of the sheet. So the second route can be considered to involve a longer trajectory due to the large deformation of the 'space' produced by node 4.

Example 19.2

Consider the following hypothetical situation. It is necessary to transport a dangerous substance from the Youngstown–Warren Regional Airport located in Northeast Ohio to the Elko Mini–JC Harris Field Airport in Nevada, USA. The transportation should be carried out using only the available commercial routes connecting airports in the USA.

The shortest path connecting both airports is: Youngstown–Warren (1) to Akron–Canton Regional (8) to James M. Cox Dayton International (29) to Dallas/Fort Worth International (118) and from there finally to Elko (17). In parenthesis we have given the number of airports to which the corresponding one is connected. The Dallas/Fort Worth International airport is the second most connected in the USA airport transportation network after Chicago O'Hare international, which has 139 connections. Consequently, if an unfortunate accident occurs during the transportation that necessitates the closure of Dallas/Fort Worth airport, it will have catastrophic consequences for the normal operation of the USA transportation network.

We can calculate the minimal communicability distance path connecting the two airports targeted in this example. The route with the minimum communicability distance is: Youngstown (1) to Akron (8) to Dayton (29) to Louisville (18) to Birmingham (17) to William P. Hobby (27) to Tulsa (16) to Elko (17). As can be seen, this route does not include any major hub of the network, e.g., none of the airports involved in this route is among the top 40 most connected ones. Consequently, an unfortunate accident in any airport in this route will not produce such catastrophic damage as the one produced by the shortest path route. The shortest path route involves four steps and that for the communicability distance involves seven steps, but the first has a sum of communicability distances equal to 3.30×10^8, while the second has a sum of communicability distances less than half of the first, 1.56×10^8.

FURTHER READING

Estrada, E., Complex networks in the Euclidean space of communicability distances, Physics Review E 85:066122, 2012.

Estrada, E., Hatano, N., and Benzi, M., The physics of communicability in complex networks, Physics Reports 514:89–119, 2012.

Statistical Physics Analogies

20

In this chapter

We introduce the basic concepts of thermodynamics and of statistical mechanics. We study the micro- and macro-canonical ensembles for an isolated network and a network in a thermal reservoir, respectively. We derive the fundamental formulae for entropy, Helmholtz, and Gibbs free energies for classical and quantum systems. We make an interpretation of the concept of temperature as applied to network sciences.

20.1 Motivation

In Chapters 8 and 16 we introduced classical and quantum mechanical analogies for studying complex networks. Here, we develop analogies from statistical mechanics to use in network sciences. The term 'statistical mechanics' was introduced by Gibbs in the nineteenth century as a means to emphasize the necessity of using statistical tools in describing macroscopic physical systems, since it is impossible to deduce the properties of such systems by analysing the mechanical properties of its individual constituents. It is possible to obtain some macroscopic measurements of the system and the goal of statistical mechanics is to use statistical methods to connect these macroscopic properties of the system with the microscopic structure and dynamics occurring in them.

A fundamental concept of both thermodynamics and statistical mechanics is that of entropy. This concept has proved useful in areas far removed from thermal physics and it is now ubiquitous in many fields of physical, biological, and social sciences. In this chapter we will prepare the terrain for understanding entropy and its implications for studying networks. We will consider briefly the following fundamental questions: What is the physical meaning of entropy? How is it connected to the information content of a network? What is the network temperature? What is the connection between network entropy and other thermodynamic properties?

20.2   Thermodynamics in a nutshell

In thermodynamics, we consider a *system* to be a part of the physical universe to which we direct our study. In our particular case the system is a network, or, as we will see later, in Section 20.4 a network and a thermal reservoir. If this system is considered according to the laws of classical mechanics (see Chapter 8), a state of the system corresponds to its complete description by means of the position and momentums of the particles forming the system at a given time. In quantum mechanics, a state is defined according to the first postulate of quantum mechanics (see Chapter 16). That is, the quantum state is completely specified by the wave function $\psi(x, t)$. A *microstate* of the system is the state in which all the parameters of the constituent parts of the system are specified. On the other hand, a *macrostate* is a state of the system in which the distribution of particles over the energy levels is specified.

Let us start by considering a network in which we have arbitrarily divided the set of nodes into two groups (see Figure 20.1). First, we assume that the network has no interactions with any other system i.e. it is isolated. A system is *isolated* if it is unable to exchange energy and matter with the exterior of the system. We also consider that the state in which the network exists does not change in time. When the probability of finding the system in a particular microstate does not change with time we say that the system is in *equilibrium*.

Label the set of black nodes V_1 and the white nodes V_2, and let $n_1 = |V_1|$ and $n_2 = |V_2|$. Denote by U_1 and U_2 the internal energies of the two parts of the network. In Figure 20.1, a wall artificially divides the nodes marked in white from those marked in black. Let us remove this wall. The total number of nodes is now $n = n_1 + n_2$ and (assuming that there is no energy of interaction between the two parts) the total internal energy is $U = U_1 + U_2$. These properties which can be obtained as the sum of the values they have in the single systems are called *extensive*. The *fundamental postulate of thermodynamics* states that it should

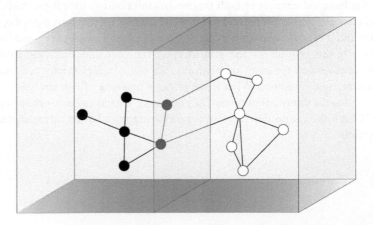

Figure 20.1 *A network with two regions isolated*

be possible to characterize the state of a thermodynamic system by specifying the values of a series of extensive variables $(X_0, \ldots, X_k)$.

We wish to exploit some of the fundamental laws and principles of thermodynamics. We start with the so-called laws of thermodynamics.

Zeroth law: If the systems A and B are in thermal equilibrium with a third system C, then A and B must also be in thermal equilibrium.

First law: The internal energy U of an isolated system is constant. The change of internal energy produced by a process that causes the system to absorb heat Q and do work W is given by $\Delta U = Q - W$.

Second law: For an isolated system, the entropy S is a state function. No process taking place in the system can decrease the entropy. That is, $\Delta S \geq 0$. If the system absorbs an infinitesimal amount of heat δQ (the system is not isolated), the entropy changes by

$$dS = \frac{\delta Q}{T}, \tag{20.1}$$

where T is the temperature of the system. Notice that while dS is an exact differential, δQ is not.

Third law: $\lim_{T \to 0} S = S_0$, where S_0 is a constant independent of all parameters of the system.

In a system in which the work done is purely mechanical, we can combine the first and the second laws and write

$$dU = TdS - \delta W = TdS - pdV,$$

and so by the product rule,

$$d(U + pV) = TdS + Vdp.$$

The quantity $H = U + pV$ is known as *enthalpy*. This gives a thermodynamic definition of the temperature as

$$T = \left(\frac{\partial H}{\partial S} \right)_p. \tag{20.2}$$

An important relation between the enthalpy, entropy, and temperature is given by the Gibbs free energy

$$F = H - TS. \tag{20.3}$$

20.3   Micro-canonical ensembles

Consider an isolated system and imagine a new system formed from an assembly of copies of this first system. This is the so-called *micro-canonical ensemble*. From the fundamental postulate of statistical mechanics, the entropy of the system is given by

$$S = k_B \ln N_\Gamma, \tag{20.4}$$

where N_Γ is the number of microscopic states accessible to the system, given by the volume of the phase space in which the observables of the system have the specified values of the extensive variables; and k_B is the *Boltzmann constant*. This gives a physical interpretation of entropy as a measure of the dispersion of the distribution of microscopic states in the system.

Recall from Chapter 8 that for the classical simple harmonic oscillator (SHO) we can write

$$E = \frac{p^2}{2m} + \frac{m\omega^2 x^2}{2}, \tag{20.5}$$

which can be written as the equation of an ellipse

$$1 = \frac{x^2}{x_{max}^2} + \frac{p^2}{p_{max}^2}. \tag{20.6}$$

where

$$p_{max} = \sqrt{2mE}, \tag{20.7}$$

$$x_{max} = \sqrt{\frac{2E}{m\omega^2}} \tag{20.8}$$

and the 'surfaces' of constant energy are described by ellipses with these axes. In this case, the volume of the phase space is the area of the ellipse

$$\pi p_{max} x_{max} = \pi \sqrt{\frac{4E^2}{\omega^2}} = \frac{2\pi E}{\omega}. \tag{20.9}$$

Suppose that the energy can only be measured with a degree of uncertainty ΔE. The area of the accessible phase space is given by

$$\Gamma_a = \frac{2\pi E + \Delta E}{\omega} - \frac{2\pi E}{\omega} = 2\pi \frac{\Delta E}{\omega}. \tag{20.10}$$

Thus, the number of microscopic states accessible to the system is obtained by dividing Γ_a by h_0, the size of a cell, in which the phase space is partitioned

$$N_\Gamma = 2\pi \frac{\Delta E}{h_0 \omega}, \tag{20.11}$$

or assuming the smallest possible length of the cell we obtain

$$N_\Gamma = \frac{\Delta E}{\hbar\omega}.$$

(20.12)

We can now obtain the value of the entropy for the SHO by using $S = k_B \ln N_\Gamma$. However, because N_Γ uses arbitrary units, we can only define S up to an arbitrary additive constant. We write

$$S = k_B \ln \frac{\Gamma_a}{\Gamma_0},$$

(20.13)

where Γ_0 has dimensions $(pq)^{dn}$ with d being the dimension of the space and n the number of particles. Finally, the entropy of the classical SHO is given by

$$S = k_B \left[\ln\left(\frac{E}{\hbar\omega}\right) + 1 \right].$$

(20.14)

20.4 The canonical ensemble

Now suppose that a network is submerged into a thermal bath as illustrated in Figure 20.2. A thermal bath or reservoir is a system with dimensions much larger than the network and the energy of interaction between the network and the reservoir is negligible. That is, if E_R is the energy of the reservoir and E is the energy of the network then $E_R \gg E$ and the total energy of the combined network plus reservoir, $E_{Tot} = E + E_R$, is also much larger than E. The ensemble representing a system which is in thermal contact with a heat reservoir, which is characterized by a temperature T is known as the *canonical ensemble*.[1]

The probability distribution for the energy in the reservoir plus network system is given by

$$P(E) = \frac{\Gamma(E)\Gamma_R(E_R)}{\Gamma_T(E_T)} = \frac{\Gamma(E)\Gamma_R(E_T - E)}{\Gamma_T(E_T)},$$

(20.15)

where $\Gamma_R(E_T - E)$ is the volume of phase space accessible to the reservoir.

[1] In the macrocanonical ensemble (not studied here) the system can interchange heat and matter with the exterior.

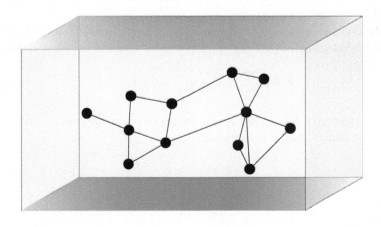

Figure 20.2 *A network submerged into a thermal bath*

By taking the logarithm of this expression we get

$$\ln P(E) = \ln \Gamma(E) + \ln \Gamma_R(E_T - E) - \ln \Gamma_T(E_T). \qquad (20.16)$$

Due to the fact that $E_T \gg E$, it is reasonable to assume that $\ln \Gamma_R(E_T - E)$ can be expanded in a Taylor series in powers of E in which we can neglect all terms of order higher than one,

$$\ln \Gamma_R(E_T - E) = \ln \Gamma_R(E_T) - E \left. \frac{\partial \ln \Gamma_R(E)}{\partial E} \right|_{E_T} + \cdots . \qquad (20.17)$$

Now, because $S = k_B \ln \Gamma$ we have

$$\left. \frac{\partial \ln \Gamma_R(E)}{\partial E} \right|_{E_T} = \frac{1}{k_B} \left. \frac{\partial S_R(E)}{\partial E} \right|_{E_T} = \frac{1}{k_B T} = \beta, \qquad (20.18)$$

where β is the so-called inverse temperature. So we can write (20.16) as

$$\ln P(E) = \ln \Gamma_R(E_T) - \ln \Gamma_T(E_T) + \ln \Gamma(E) - \beta E. \qquad (20.19)$$

The first two terms in (20.19) are constant and if we define the constant

$$Z = \frac{\Gamma_T(E_T)}{\Gamma_R(E_T)} \qquad (20.20)$$

then

$$\ln P(E) = \ln \Gamma(E) - \beta E - \ln Z \qquad (20.21)$$

or

$$P(E) = \frac{1}{Z} \Gamma(E) \exp(-\beta E). \qquad (20.22)$$

The normalization constant Z plays a very important role in statistical mechanics. It is known as the partition function of the system and can be expressed as

$$Z = \int \frac{d\Gamma}{\Gamma_0} \exp(-\beta \mathcal{H}) = \mathrm{tr}(\exp(-\beta \mathcal{H})). \qquad (20.23)$$

The Helmholtz free energy of the system can now be obtained from the partition function using the expression

$$F = -\beta^{-1} \ln Z. \qquad (20.24)$$

For instance, for the classical SHO the partition function is written as

$$Z = \int \frac{dx\,dp}{h_0} \exp\left[-\beta\left(\frac{p^2}{2m} + \frac{m\omega^2 x^2}{2}\right)\right],$$ (20.25)

which after integration gives

$$Z = \frac{k_B T}{\hbar\omega},$$ (20.26)

so

$$F = -k_B T \ln\left(\frac{k_B T}{\hbar\omega}\right) = k_B T \ln\left(\frac{\hbar\omega}{\beta}\right),$$ (20.27)

and

$$S = -\left(\frac{\partial F}{\partial T}\right)_V = k_B\left[\ln\left(\frac{k_B T}{\hbar\omega}\right) + 1\right].$$ (20.28)

Problem 20.1
Find expressions for the partition function, entropy, Helmholtz, and Gibbs free energies of the simple quantum harmonic oscillator.

For the quantum SHO we have seen that

$$\widehat{\mathcal{H}} = -\hbar\omega\left(\widehat{N} + \frac{1}{2}\right).$$

Thus the partition function is

$$Z = \mathrm{tr}(\exp(-\beta\widehat{\mathcal{H}})) = \mathrm{tr}\left(\exp\left[-\hbar\omega\left(\widehat{N} + \frac{1}{2}\right)\right]\right),$$

which can be written as

$$Z = \sum_{j=0}^{\infty} \exp\left[-\beta\hbar\omega\left(j + \frac{1}{2}\right)\right] = \exp\left[-\frac{\beta\hbar\omega}{2}\right]\sum_{j=0}^{\infty} \exp[-\beta\hbar\omega j].$$

Then,

$$Z = \frac{\exp\left[-\frac{\beta\hbar\omega}{2}\right]}{1 - \exp[-\beta\hbar\omega]} = \frac{1}{\exp\left(\frac{\beta\hbar\omega}{2}\right) - \exp\left(-\frac{\beta\hbar\omega}{2}\right)}$$

$$= \frac{1}{2\sinh\left(\frac{\beta\hbar\omega}{2}\right)} = \frac{1}{2}\mathrm{csch}\left(\frac{\beta\hbar\omega}{2}\right).$$

Hence, from (20.27), the Helmholtz free energy of the system is

$$F = \beta^{-1} \ln \left[2 \sinh \left(\frac{\beta \hbar \omega}{2} \right) \right],$$

and by (20.28) the entropy is

$$S = -\left(\frac{\partial F}{\partial T} \right) = \frac{\hbar \omega}{2T} \coth \left(\frac{\beta \hbar \omega}{2} \right) - k_B \ln \left[2 \sinh \left(-\frac{\beta \hbar \omega}{2} \right) \right].$$

Finally, from (20.3),

$$H = \frac{\hbar \omega}{2} \coth \left(\frac{\beta \hbar \omega}{2} \right).$$

There are alternative expressions for the entropy of a system in the quantum canonical ensemble and we now derive one. The probability of finding the system in a quantum state with energy E_j is given by

$$p_j = \frac{1}{Z} \exp(-\beta E_j), \tag{20.29}$$

which implies that

$$\ln p_j = -(\ln Z + \beta E_j), \tag{20.30}$$

with the condition that $\displaystyle\sum_{j=1}^{n} p_j = \frac{1}{Z} \sum_{j=1}^{n} \exp(-\beta E_j) = 1.$

Then from (20.27)

$$S = -\left(\frac{\partial F}{\partial T} \right)_V = k_B \ln Z + k_B T \left(-\frac{1}{k_B T^2} \right) \frac{\partial}{\partial \beta} \ln Z$$

$$= k_B \left[\ln Z - \frac{\beta}{Z} \sum_j (-E_j) \exp(-\beta E_j) \right] = k_B \left[\ln Z \sum_j p_j - \frac{\beta}{Z} \sum_j (-E_j) \exp(-\beta E_j) \right]$$

$$= k_B \sum_j p_j (\ln Z + \beta E_j) = -k_B \sum_j p_j \ln p_j.$$

This is a beautiful result in which we see that

$$S = -k_B \sum_j p_j \ln p_j, \tag{20.31}$$

[2] You may be familiar with the Shannon formula $S = -\sum_j p_j \ln p_j$ which can be applied to any probability distribution.

which connects the information content of a system[2] with the thermodynamics of the system.

20.5 The temperature in network theory

It seems *a priori* very artificial to consider that we submerge a real-world network into a thermal bath. But we argue that in reality, real-world networks are already submerged into such 'thermal baths'. Think, for instance, of our network of social relations. It is contained in a reservoir of the global social, economic, and political atmospheres of the society as a whole. Changes in these atmospheres can change the strength of our social ties. A protein–protein interaction network is contained in the cell reservoir and subject to any variation in the physical parameters (pH, temperature, concentration of nutrients, etc) in that cell. A network of corporations and their business relationships is embedded into the global economic climate at the moment the network is analysed. Thus the inverse temperature β can play the role of the external stress to which any network is submitted due to any change in the environment in which the network is submerged.

Consider the network illustrated in Figure 20.3, which has been submerged into a thermal bath with inverse temperature β. After equilibration, every link of the network is weighted by β. If the network is already weighted then every link weight is multiplied by β.

This means that as the temperature tends to infinity, $\beta \to 0$ and all links tend to have zero weights. In other words, the network is fully disconnected. This resembles the situation of very extreme stress which makes the network totally dysfunctional. When the temperature tends to zero, $\beta \to \infty$ and all links have infinite weights, indicating an infinite strengthening of the connections between bonded nodes.

Problem 20.2

Consider a tight-binding model for a network with parameters $\tilde{\alpha} = 0$ and $\tilde{\beta} = -1$. Define the partition function, entropy, Helmholtz, and Gibbs free energies for a network.

The Hamiltonian for the tight-binding model is $\widehat{\mathcal{H}} = \tilde{\alpha} I + \tilde{\beta} A$ and the partition function of the network with $\tilde{\alpha} = 0$ and $\tilde{\beta} = -1$ is

$$Z = \text{tr}(\exp(-\beta\widehat{\mathcal{H}})) = \text{tr}(\exp(\beta A)).$$

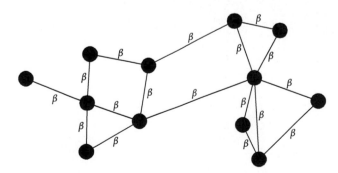

Figure 20.3 *Effects of equilibration of a network in a thermal bath with inverse temperature β*

In terms of the eigenvalues of the adjacency matrix we have

$$Z = \sum_{j=1}^{n} \exp(\beta\lambda_j).$$

Notice that the partition function of the network is just the sum of the subgraph centralities of the network. The index $\text{tr}(\exp(\beta A))$ is usually called the Estrada index of the network. Using (20.31) we can obtain the entropy of the network in the canonical ensemble. Since $E_j = -\lambda_j$, we have $p_j = \exp(\beta\lambda_j)/Z$ and so,

$$S = -k_B \sum_j p_j(\beta\lambda j - \ln Z) = -\frac{1}{T}\sum_j \lambda_j p_j + k_B \ln Z \sum_j p_j = -\frac{1}{T}\sum_j \lambda_j p_j + k_B \ln Z,$$
(20.32)

which can be rearranged to give

$$-\beta^{-1}\ln Z = -\sum_j \lambda_j p_j - TS.$$

Using the expression $F = H - TS$ we obtain $H = -\sum_j \lambda_j p_j$ and $F = -\beta^{-1}\ln Z$.

....

FURTHER READING

Albert, R. and Barabási, A., Statistical mechanics of complex networks, Reviews of Modern Physics 74:47–97, 2002.

Park, J. and Newmann, M.E.J., Statistical mechanics of networks. Physical Review E 70:066117, 2004.

Petiti, L., *Statistical Mechanics in a Nutshell*, Princeton University Press, 2011.

Communities in Networks

In this chapter

We study the organization of nodes into communities in complex networks. Communities are groups of nodes more densely connected amongst themselves than with the rest of the nodes of the network. We study some of the methods that aim to find such structures in networks. We finish with a method to detect anti-communities (bipartitions) in networks.

21.1   Motivation

In real-world networks, nodes frequently group together forming densely connected clusters which are poorly connected with other parts of the network. Clusters, also known as communities in network theory, may form for many reasons. For instance, we all belong to clusters formed by our friends and relatives. Inside these groups one can find a relatively high density of ties but in many cases these are poorly connected to others groups in society. Clusters can also be formed due to similarities among the nodes. For instance, groups of proteins with similar functions in a protein–protein interaction network may be more densely connected with each other than with proteins which have different functions. In this chapter, we study how to find these communities of nodes based on the information provided by the topological structure of the network.

Examples 21.1

(i) In Figure 21.1 we illustrate two well-known cases of networks with communities. The first corresponds to the friendship ties among individuals in a karate club in a USA university. At some point in time, the members of this social network were polarized into two different factions due to an argument between the instructor and the president. These two factions, represented in Figure 21.1(a) in different colours, act as cohesive groups which can be considered as independent entities in the network. Another example is provided in Figure 21.1(b).

continued

228   *Communities in Networks*

Examples 21.1 *continued*

The network represents 62 bottlenose dolphins living in Doubtful Sound, New Zealand. Links are drawn between two animals if they are seen together more frequently than expected at random. These dolphins split into groups after one particular dolphin moved away for a period of time.

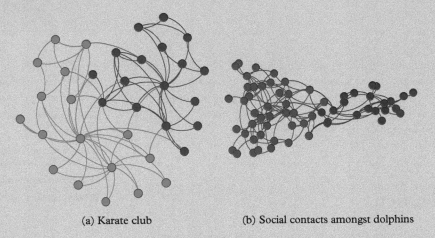

(a) Karate club   (b) Social contacts amongst dolphins

Figure 21.1 *Two networks with known structural communities*

(ii) Political polarization in the USA leads to a number of clustered networks. We illustrate this with an example based on political literature. Figure 21.2 represents a network of books on US politics published around the time of the 2004 presidential election and sold on Amazon.com. There is a link between two books if they share several common purchasers. As can be seen, there is a clear congregation into two main communities. These represent the purchases of consumers of conservative literature on one side and of liberal literature on the other.

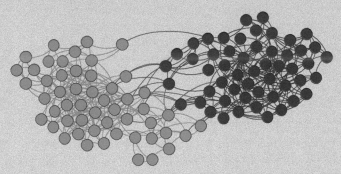

Figure 21.2 *Political books on US politics published around 2004. Nodes coloured according to the political leaning of the book*

In these examples, the clusters were induced by empirical evidence. The question that arises is whether we can find such partitions in networks without any other information than that provided by the topological structure of the network. When looking at the graph Laplacian we saw that such an approach is viable via the Fiedler vector. We will return to that subject in Section 21.3, but we will consider several other approaches, too.

21.2 Basic concepts of communities

Before attempting to detect communities, we will define some useful terms and quantities. Given a subgraph $G_1(C, E_1) \subseteq G(V, E)$ with $n_C = |C|$ nodes, we define the internal and external degrees to be the quantities

$$k_i^{int} = \sum_{j \in C} a_{ij}, \quad k_i^{ext} = \sum_{j \in \overline{C}} a_{ij},$$

respectively, where $\overline{C}$ is the complement of C and A is the adjacency matrix of G. The number of links which connect nodes internally in the subgraph is given by

$$m_C = \frac{1}{2} \sum_{i \in C} k_i^{int},$$

and the number of links connecting C and $\overline{C}$ is

$$m_{C-\overline{C}} = \sum_{i \in C} k_i^{ext}.$$

The number of links that connect nodes in C to $\overline{C}$ is called the boundary of C (and is often written ∂C). Finally, we define intra-cluster density. A network with n nodes can have at most $n(n-1)/2$ edges and we can define the density of a network with m edges and n nodes to be

$$\delta(G) = \frac{2m}{n(n-1)}.$$

Similarly the *intra-cluster density* is

$$\delta_{int}(C) = \frac{m_C}{n_C(n_C - 1)/2} = \frac{\sum_{i \in C} k_i^{int}}{n_C(n_C - 1)}.$$

There can be at most $n_1 n_2$ edges between two groups of nodes of size n_1 and n_2. So we define the *inter-cluster density* to be

$$\delta_{ext}(C) = \frac{m_{C-\overline{C}}}{n_C(n - n_C)} = \frac{\sum_{i \in C} k_i^{ext}}{n_C(n - n_C)}.$$

Example 21.2

We can calculate the quantities defined above for the networks in Figure 21.1.

In the karate club network, the values of the internal densities of the communities C_1 (followers of the instructor) and C_2 (followers of the president) are $\delta_{int}(C_1) = 0.26$ and $\delta_{int}(C_2) = 0.24$, respectively, and the inter-cluster density is $\delta_{ext}(C_1) = \delta_{ext}(C_2) = 0.035$. The total density of this network is $\delta(G) = 0.14$.

For the network of bottlenose dolphins, the cluster represented in blue has $\delta_{int}(C_1) = 0.26$, while the one represented in red has $\delta_{int}(C_2) = 0.14$ and the external density is $\delta_{ext}(C_1) = \delta_{ext}(C_2) = 0.007$, while the total density of the network is $\delta(G) = 0.08$.

For these clusters found experimentally the internal density of every cluster is significantly larger than its external density, and of the total density of the network. Another observation is that there is at least one path between every pair of nodes in a community linked together by edges and nodes in the same community. That is, communities are *internally connected*. We can loosely define a cluster, or community as follows.

Definition 21.1 *A cluster (or community) of a network is an internally connected set of nodes for which the internal density is significantly larger than the external one.*

This principle is the basis for the construction of networks with explicit clusters, which are known as benchmark graphs. Although our two examples relate to social networks, clustering is important in applications involving many other types of network, too.

21.3 Network partition methods

We now consider the problem of partitioning a network $G(V, E)$ into p disjoint sets of nodes such that the following properties hold.

- $\bigcup_{i=1}^{p} V_i = V$ and $V_i \cap V_j = \varnothing$ for $i \neq j$.
- The number of links crossing between subsets (cut-set size or boundary) is minimized.
- $|V_i| \approx n/p$ for $1 \leq i \leq p$.

The process can be generalized to weighted networks where the cut-set is defined as the sum of the weights of the links crossing subsets and the final condition is rewritten as $W_i \approx W/p$, where W_i is the sum of the weights of the nodes in V_i and W is the total sum of weights in the whole network. Such a weighted partition is called *balanced*.

We now analyse some of the most popular partitioning algorithms.

21.3.1　Local improvement methods

Local improvement methods are amongst the earliest proposed partitioning algorithms and the archetypal method is due to Kernighan and Lin. In general, these methods take a partition of the network as their input; the most simple case is by taking a bisection. They then try to decrease the cut-set with a local search approach. The generation of the initial partition can be carried out using any popular bisection method, or can simply be generated at random. In the latter case, several random realizations are needed and the one with minimum cut size is selected as the input. For the sake of simplicity we are going to consider only unweighted networks. Given a bisection of G into subgraphs $G_1(V_1, E_1)$ and $G_2(V_2, E_2)$ we define an internal and external weight for each node by splitting its degree so W_i^{int} is the number of edges from i to nodes in the same partition and W_i^{ext} the number which leave. The cut size is given by

$$C(V_1, V_2) = \frac{1}{2} \sum_{i \in V} W_i^{ext}.$$

If we swap node i from one set to another we can measure the gain in the cut size by $g_i = W_i^{ext} - W_i^{int}$, a positive gain indicating an improvement.

Example 21.3

Let us consider the network illustrated in Figure 21.3. We will return to this network repeatedly in this chapter and we name it G_{clus}. A random partition of the nodes into subsets of equal size is represented by the dotted line.

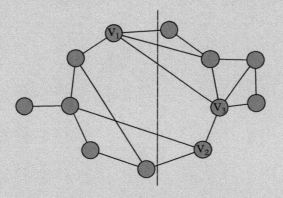

Figure 21.3　*A random partition of a network*

continued

Example 21.3 *continued*

Note that $g_{v_1} = W_{v_1}^{ext} - W_{v_1}^{int} = 3 - 1 = 2$. The gains for nodes v_2 and v_3 are 1 and -3, respectively. The cut size for this partition is $C = 5$.

We reduce the cut size by moving nodes with positive gain across to the other side and the reduction in C is equal to the gain. Moving node v_1 reduces C to 3, as seen in Figure 21.4.

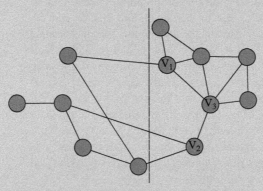

Figure 21.4 *An improvement to the partition in Figure 21.3*

Moving v_1 unbalances the number of nodes in each partition. Rather than simply analysing the gain obtained by moving one node from one partition to another, we need to quantify the gain produced by swapping two nodes in opposite partitions. If $v_1 \in V_1$ and $v_2 \in V_2$ then the gain of interchanging is given by

$$g(v_1, v_2) = \begin{cases} g(v_1) + g(v_2) - 2, & v_1 \sim v_2, \\ g(v_1) + g(v_2), & \text{otherwise.} \end{cases}$$

In our example, since v_1 and v_2 are not adjacent, $g(v_1, v_2) = 3$ and the cut size is reduced from 5 to 2. The new partition is shown in Figure 21.5.

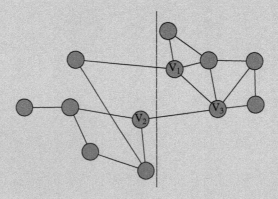

Figure 21.5 *An even better partition of Figure 21.3*

This node swapping process forms the basis for the *Kernighan–Lin algorithm*. We start with a balanced bisection $\{V_1, V_2\}$ and we compute the cut size, $C_0(V_1, V_2)$, of the bisection. Then for $k = 1, 2, \ldots, r$, where $r = \min(|V_1|, |V_2|)$ we carry out the following steps.

- Find the pair of nodes $v_1 \in V_1$ and $v_2 \in V_2$ which give the biggest value of $g(v_1, v_2)$ (which may be negative).
- Label these nodes v_1^k and v_2^k.
- For any node u adjacent to either v_1^k or v_2^k we update the value of $g(u)$.
- Calculate

$$C_k(V_1, V_2) = C_{k-1}(V_1, V_2) - g(v_1^k, v_2^k), \quad k = 1, \ldots, r,$$

and note $C_j(V_1, V_2)$, the minimum of all these values.
- Update the partitions by moving the sets

$$\{v_1^1, v_1^2, \ldots, v_1^j\} \quad \text{and} \quad \{v_2^1, v_2^2, \ldots, v_2^j\}$$

between V_1 and V_2.
- Repeat the whole process until no further improvement is achieved for the cut size.

The algorithm requires a time proportional to the third power of the number of nodes in the network, $O(n^3)$, but improvements can be made to reduce this cost significantly. In brief, these changes improve the process for swapping nodes; use a fixed number of iterations; and only evaluate gain for nodes close to the partition boundary. We introduce this algorithm here not because it is used today for detecting communities in networks but because it helps us to understand the intuition behind the partitioning methods for detecting communities.

21.3.2   Spectral partitioning

We have seen that the Fiedler vector can be used to partition a connected network into two. This technique, which stems from the use of an eigenvector, is an example of *spectral partitioning*. In this section we will develop these ideas and see that it is possible to use spectral information to partition a network into several pieces.

Recall that the Fiedler vector is the eigenvector, $\mathbf{x}$, associated with the smallest nontrivial eigenvalue of the graph Laplacian. We associate the ith component of this vector with the ith node and pick a cut-off value r. The nodes are then split according to whether x_i is bigger or smaller than r. The choice $r = 0$ is often good in practice.

In 1988, Powers showed that eigenvectors of the adjacency matrix can also be used to find partitions in a network. The idea behind these methods is that the second largest eigenvector q_2^A has both positive and negative components, allowing a partition of the network according to the sign pattern of this eigenvector.

Example 21.4

The eigenvector associated with the second smallest eigenvalue of the Laplacian matrix of G_{clus} is

$$q_2^L = [-0.30 \quad -0.13 \quad 0.18 \quad 0.29 \quad 0.28 \quad 0.32 \quad 0.34 \quad 0.22 \quad -0.12 \quad -0.22$$
$$-0.33 \quad -0.51]^T.$$

Similarly, the eigenvector associated with the second largest eigenvalue of the adjacency matrix of this network is

$$q_2^A = [-.50 \quad -0.34 \quad 0.08 \quad 0.13 \quad 0.24 \quad 0.23 \quad 0.16 \quad 0.16 \quad -0.31$$
$$-0.42 \quad -0.37 \quad -0.20]^T.$$

Both eigenvectors produce the same bipartition of this network as the one produced by the Kernighan–Lin algorithm.

Sometimes we might expect there to be more than two communities within a network. Spectral techniques can easily be extended by using extra eigenvectors. We describe the process with respect to the Laplacian matrix, but there is an analogous approach for the adjacency matrix.

Suppose, then, that we split the network into two according to the signs of the Fiedler vector. We now look at the eigenvector, y, associated with the next smallest eigenvalue. For each of our original clusters we look at the signs of the elements in the corresponding positions of y and split these clusters into those with positive and those with negative components. We have now produced four clusters and we can carry on dividing by looking at further eigenvectors. If we are convinced a cluster should be divided no further, then we can only apply the information provided by these eigenvectors to sections of the network. If we go too far down this route there is less and less mathematical justification, but applied carefully it can provide convincing clusters.

Example 21.5

In Figure 21.6 we show the clusters induced in the karate club by using two eigenvectors. In (a) we have used the Laplacian matrix and in (b) the adjacency matrix. Notice that in picture (b) not all of the clusters are connected.

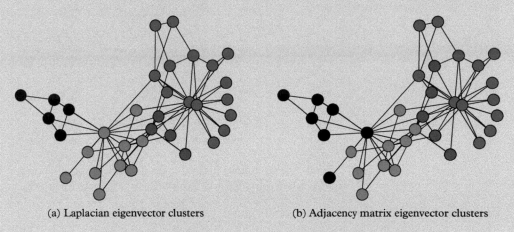

(a) Laplacian eigenvector clusters   (b) Adjacency matrix eigenvector clusters

Figure 21.6 *Two partitions of the Zachary karate club induced by multiple eigenvectors*

21.4   Clustering by centrality

Any of our centrality measures can be used to divide a network into clusters. The idea is to identify those links which are central for the inter-cluster communication. By removing them, we can find the best partition of the network into communities. The first such algorithm, proposed by Girvan and Newman, used edge betweenness centrality. This is a simple extension of the idea of node betweenness centrality that we have seen previously: the betweenness centrality of an edge a is found by dividing the number of shortest paths between two nodes that pass through a by the total number of such shortest paths. The expectation is that links connecting nodes which are located in different communities display the highest edge betweenness.

Example 21.6

In Figure 21.7 we highlight the two links with highest edge betweenness centrality for G_{clus}.

continued

Example 21.6 *continued*

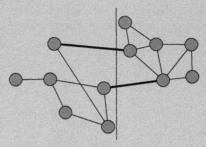

Figure 21.7 *Selecting links according to highest betweenness centrality*

Removing these links partitions the network in the same way as by using the Kernighan–Lin approach but does not require an initial bisection.

The *Girvan–Newman* algorithm for partitioning by edge betweenness can be described as follows.

- Calculate the edge betweenness centrality for all links in the network.
- Remove all links with the largest edge betweenness.
- Repeat these steps on the new network and continue until all links have been removed.

We can use the information provided by this algorithm to provide a hierarchical picture for analysing the community structure of the network. We can picture this information in a dendrogram which we build from the bottom to the top.

In a network with n nodes we start by drawing n points to representing these nodes. Then we find $n-1$ clusters by joining one pair of nodes and leaving the rest isolated. We add links one at a time, pairing up clusters until we have brought the network back together. We link clusters by replacing edges in the network in the reverse order of their removal in the Girvan–Newman algorithm.

A significant characteristic of this method is that it gives a range of possible partitions of the network instead of a fixed one. The authors of the algorithm state that 'it is up to the user to decide which of the many divisions represented is most useful for their purposes'. For complex networks arising in real-world situations it is preferable to automate this decision making process. This requires us to develop a quality criterion in order to select one partition over the others. We will return to this idea shortly.

Example 21.7

The dendrogram for G_{clus} is illustrated in Figure 21.8. The top dotted line indicates the division of the network into two communities, each formed by six nodes. The second dotted line indicates a division into four communities having 3, 3, 5, and 1 nodes, respectively.

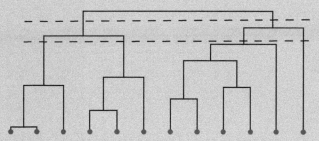

Figure 21.8 *Dendrogram obtained by the Girvan–Newman algorithm*

21.5 Modularity

Girvan and Newman proposed *modularity* as a measure of quality of clusters. They start with the assumption that a cluster is a structural element of a network that has been formed in a far from random process. If we consider the actual density of links in a community, it should be significantly larger than the density we would expect if the links in the network were formed by a random process.

Definition 21.2 *Let $G(V,E)$ be a network of n nodes and m edges with adjacency matrix A and suppose we have divided the nodes into n_C clusters $V_1, V_2, \ldots, V_{n_C}$. Define s_{ir} to equal 1 if node i is in cluster r and 0 otherwise. Then the modularity of the partitioning is given by*

$$Q = \frac{1}{4m} \sum_{r=1}^{n_C} \sum_{i,j=1}^{n} \left(a_{ij} - \frac{k_i k_j}{2m} \right) s_{ir} s_{jr},$$

where k_i is the degree of node i.

Modularity can be interpreted as the sum over all partitions of the difference between the fraction of links inside each partition and the expected fraction, by considering a random network with the same degree for each node, giving

$$Q = \sum_{k=1}^{n_C} \left[\frac{|E_k|}{m} - \frac{1}{4m^2} \left(\sum_{j \in V_k} k_j \right)^2 \right], \tag{21.1}$$

where $|E_k|$ is the number of links between nodes in the kth partition of the network. If the number of intra-cluster links is no bigger than the expected value for a random network then $Q = 0$. The maximum modularity is one and the more positive the value of Q, the more convincing is the community structure.

Example 21.8

We label the two communities in G_{clus} found with the Kernighan–Lin algorithm as C_1 and C_2 as illustrated in Figure 21.9.

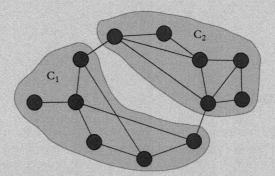

Figure 21.9 *Clusters induced by the Kernighan–Lin algorithm*

The number of edges and sum of degree for the communities are

$$|E_{C_1}| = 7, \quad \sum_{j \in V_{C_1}} k_j = 16, \quad |E_{C_2}| = 9, \quad \sum_{j \in V_{C_2}} k_j = 20,$$

and the total number of edges is $m = 18$. From (21.1),

$$Q = \underbrace{\frac{7}{18} - \left(\frac{16}{36}\right)^2}_{C_1} + \underbrace{\frac{9}{18} - \left(\frac{20}{36}\right)^2}_{C_2} = 0.383.$$

The goal of finding this value is to compare it with the modularity of other partitions of the same network, as we will see later, in Section 21.6.

21.5.1   The problem of resolution

Example 21.9

A network is pictured in Figure 21.10 and the partitioning that maximizes Q is shown.

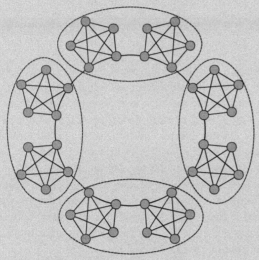

Figure 21.10 *A manifestation of the resolution problem of modularity*

In this case a more natural clustering would be achieved by splitting the network into eight communities.

The problem highlighted in the last example is a phenomenon called *resolution limit*. The modularity index can be fooled if there are some well-defined communities adjacent to bigger ones or where many small communities are circularly connected. Modularity tends to group together pairs of the small communities instead of identifying them as independent ones. This suggests that it may be better to use more than one quality measure in determining optimal partitions.

21.6   Communities based on communicability

In Chapter 19 we introduced a communicability function which accounts for the volume of information transmitted from one node to another in a network by using all possible routes between them. The concept of communicability introduces an intuitive way for finding the structure of communities in complex networks: a community is a group of nodes which have better communicability among themselves than with the rest of the nodes in the network.

We start by considering the communicability function between a pair of nodes p and q written in terms of the eigenvalues and eigenvectors of the adjacency matrix as

$$G_{pq} = \mathbf{q}_1(p)\mathbf{q}_1(q)\exp(\lambda_1) + \sum_{j=2}^{n} \mathbf{q}_j(p)\mathbf{q}_j(q)\exp(\lambda_j). \qquad (21.2)$$

Recall from Chapter 18 that if a network has a large spectral gap ($\lambda_1 \gg \lambda_2$) then we can expect it to be homogeneous, which indicates that it contains no communities or clusters due to the lack of any cut-set in the network. If this is the case, the term $\mathbf{q}_1(p)\mathbf{q}_1(q)\exp(\lambda_1)$ dominates (21.2). Whether there is a large spectral gap or not, we can interpret this term as the extent to which the network forms a single cohesive unit and use the other terms to indicate potential clusters. One approach is to make use of the signs of the elements of eigenvectors. This is illustrated in Figure 21.11 where we show the sign of the eigenvectors corresponding to the four largest eigenvalues of G_{clus}. As we have seen, the sign pattern of the second largest eigenvector of the adjacency matrix induces a partition of the network. We represent this by using two different kinds of arrows for positive and negative contributions for λ_2, λ_3, and λ_4.

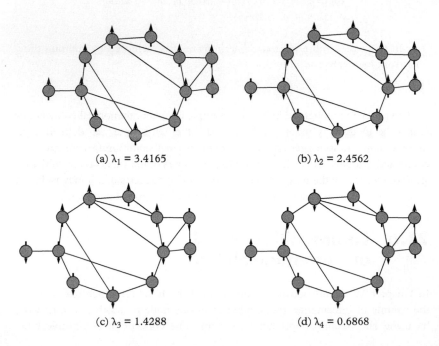

(a) $\lambda_1 = 3.4165$ (b) $\lambda_2 = 2.4562$

(c) $\lambda_3 = 1.4288$ (d) $\lambda_4 = 0.6868$

Figure 21.11 *Using eigenvector components to induce spin orientations in a network*

We can use this sign pattern of the eigenvectors to express the sum of the contributions of the non-principal eigenvalues and eigenvectors to the communicability function as

$$\sum_{j=2}^{n} \mathbf{q}_j(p)\mathbf{q}_j(q) \exp(\lambda_j) = \sum_{j=2}^{n} \mathbf{q}_j^+(p)\mathbf{q}_j^+(q) \exp(\lambda_j) + \sum_{j=2}^{n} \mathbf{q}_j^-(p)\mathbf{q}_j^-(q) \exp(\lambda_j)$$

$$- \left| \sum_{j=2}^{n} \mathbf{q}_j^+(p)\mathbf{q}_j^-(q) \exp(\lambda_j) + \sum_{j=2}^{n} \mathbf{q}_j^-(p)\mathbf{q}_j^+(q) \exp(\lambda_j) \right|,$$

$$(21.3)$$

where $\mathbf{q}_j^+(p)$ indicates that the entry corresponding to the node p of the jth eigenvector is positive.

Now, consider a cluster formed by nodes which have the same sign contribution to the communicability function. A physical interpretation is that two nodes with the same sign in one eigenvector are 'vibrating' in the same direction, so should be coupled together. On the contrary, if two nodes have different signs for the same eigenvector they are in different clusters because they are vibrating in different phases. This analogy allows us to group the two contributions of this term to the communicability in the intra- and inter-cluster communicabilities,

$$\sum_{j=2}^{n} \mathbf{q}_j(p)\mathbf{q}_j(q) \exp(\lambda_j) = G_{pq}^{\text{intra-cluster}} - \left| G_{pq}^{\text{inter-cluster}} \right|. \qquad (21.4)$$

This gives another way of defining a cluster.

Definition 21.3 *A community is a subset of nodes $C \subset V$ for which the intra-cluster communicability among every pair of nodes is larger than the inter-cluster one.*

The difference between intra- and inter-cluster communicability is given by

$$\Delta G_{pq} = G_{pq} - \mathbf{q}_1(p)\mathbf{q}_1(q) \exp(\lambda_1) = G_{pq}^{\text{intra-cluster}} - \left| G_{pq}^{\text{inter-cluster}} \right|, \qquad (21.5)$$

which means that we only need to calculate $G_{pq} - \mathbf{q}_1(p)\mathbf{q}_1(q) \exp(\lambda_1)$ in order to determine the difference between intra- and inter-cluster communicabilities. However, because of the definition of a community, we need to check ΔG_{pq} for every pair of nodes. In order to do that we can use the following algorithm.

- Form the communicability matrix $G = \exp(A)$.
- Find the largest eigenvalue λ_1 of the adjacency matrix and its corresponding eigenvector $\mathbf{q}_1$.

- Form $\Delta G = G - \mathbf{q}_1 \mathbf{q}_1^T \exp(\lambda_1)$ and then $\Delta G'$ whose entries are defined by

$$\Delta G'_{pq} = \begin{cases} 1, & \Delta G_{pq} > 0, \\ 0, & \Delta G_{pq} \le 0 \text{ or } p = q. \end{cases}$$

- Define the communicability graph to be the network with adjacency matrix $\Delta G'_{pq}$.
- Find the cliques in the communicability graph. Each clique represents a community in the network.

Example 21.10

We return to G_{clus}, labelled as in Figure 21.12.

The communicability matrix for this network is

$$G = \begin{bmatrix}
4.48 & 3.05 & 1.64 & 0.57 & 1.01 & 0.68 & 0.52 & 1.75 & 3.07 & 2.77 & 2.68 & 1.94 \\
3.05 & 3.76 & 3.01 & 1.36 & 1.96 & 1.08 & 0.72 & 2.28 & 2.35 & 2.80 & 1.84 & 0.95 \\
1.64 & 3.01 & 6.14 & 3.84 & 5.75 & 3.61 & 2.34 & 5.70 & 2.25 & 1.57 & 0.74 & 0.38 \\
0.57 & 1.36 & 3.84 & 3.31 & 3.98 & 2.18 & 1.25 & 3.24 & 0.97 & 0.55 & 0.21 & 0.11 \\
1.01 & 1.96 & 5.75 & 3.98 & 6.82 & 4.87 & 3.08 & 6.53 & 2.11 & 0.97 & 0.38 & 0.19 \\
0.68 & 1.08 & 3.61 & 2.18 & 4.87 & 4.99 & 3.54 & 5.66 & 1.81 & 0.66 & 0.25 & 0.13 \\
0.52 & 0.72 & 2.34 & 1.25 & 3.08 & 3.54 & 3.31 & 4.26 & 1.42 & 0.50 & 0.19 & 0.10 \\
1.75 & 2.28 & 5.70 & 3.24 & 6.53 & 5.66 & 4.26 & 8.04 & 3.47 & 1.67 & 0.78 & 0.39 \\
3.07 & 2.35 & 2.25 & 0.97 & 2.11 & 1.81 & 1.42 & 3.47 & 3.85 & 2.82 & 1.85 & 0.95 \\
2.77 & 2.80 & 1.57 & 0.55 & 0.97 & 0.66 & 0.50 & 1.67 & 2.82 & 3.61 & 2.44 & 0.74 \\
2.68 & 1.84 & 0.74 & 0.21 & 0.38 & 0.25 & 0.19 & 0.78 & 1.85 & 2.44 & 2.69 & 0.87 \\
1.94 & 0.95 & 0.38 & 0.11 & 0.19 & 0.13 & 0.10 & 0.39 & 0.95 & 0.74 & 0.87 & 1.71
\end{bmatrix},$$

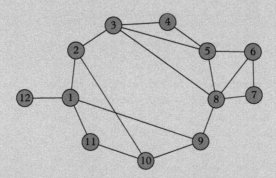

Figure 21.12 *A labelling of the network G_{clus}*

and

$$\Delta G = \begin{bmatrix} 3.49 & 1.84 & -0.58 & -0.78 & -1.37 & -1.18 & -0.81 & -0.93 & 1.73 & 1.86 & 2.12 & 1.65 \\ 1.84 & 2.29 & 0.30 & -0.28 & -0.93 & -1.19 & -0.90 & -0.97 & 0.72 & 1.70 & 1.16 & 0.59 \\ -0.58 & 0.30 & 1.16 & 0.83 & 0.43 & -0.57 & -0.64 & -0.29 & -0.75 & -0.47 & -0.51 & -0.27 \\ -0.78 & -0.28 & 0.83 & 1.49 & 0.76 & -0.35 & -0.55 & -0.38 & -0.84 & -0.68 & -0.54 & -0.29 \\ -1.37 & -0.93 & 0.43 & 0.76 & 1.15 & 0.41 & -0.10 & 0.14 & -1.09 & -1.20 & -0.95 & -0.50 \\ -1.18 & -1.19 & -0.57 & -0.35 & 0.41 & 1.48 & 1.05 & 0.63 & -0.70 & -1.04 & -0.80 & -0.42 \\ -0.81 & -0.90 & -0.64 & -0.55 & -0.10 & 1.05 & 1.53 & 0.68 & -0.37 & -0.71 & -0.55 & -0.29 \\ -0.93 & -0.97 & -0.29 & -0.38 & 0.14 & 0.63 & 0.68 & 0.84 & -0.13 & -0.78 & -0.72 & -0.39 \\ 1.73 & 0.72 & -0.75 & -0.84 & -1.09 & -0.70 & -0.37 & -0.13 & 2.05 & 1.60 & 1.10 & 0.56 \\ 1.86 & 1.70 & -0.47 & -0.68 & -1.20 & -1.04 & -0.71 & -0.78 & 1.60 & 2.78 & 1.93 & 0.47 \\ 2.12 & 1.16 & -0.51 & -0.54 & -0.95 & -0.80 & -0.55 & -0.72 & 1.10 & 1.93 & 2.38 & 0.71 \\ 1.65 & 0.59 & -0.27 & -0.29 & -0.50 & -0.42 & -0.29 & -0.39 & 0.56 & 0.47 & 0.71 & 1.63 \end{bmatrix}.$$

In Figure 21.13 we have reordered the nodes of ΔG in order to illustrate the existence of a main diagonal positive block formed by nodes 1, 2, 9, 10, 11, 12, which has a negative level of communication with the rest of the nodes. This contour plot fails to highlight the community formed by nodes 3, 4, 5, 6, 7, 8.

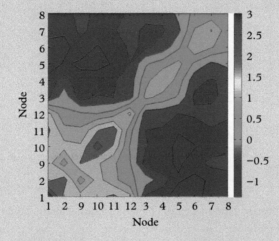

Figure 21.13 *Contour plot of ΔG for G_{clus}*

We transform ΔG into a communicability graph by replacing nonpositive values by zeroes and positive values by ones to give

continued

Example 21.10 *continued*

$$\Delta G' = \begin{bmatrix} 0 & 1 & 0 & 0 & 0 & 0 & 0 & 0 & 1 & 1 & 1 & 1 \\ 1 & 0 & 1 & 0 & 0 & 0 & 0 & 0 & 1 & 1 & 1 & 1 \\ 0 & 1 & 0 & 1 & 1 & 0 & 0 & 0 & 0 & 0 & 0 & 0 \\ 0 & 0 & 1 & 0 & 1 & 0 & 0 & 0 & 0 & 0 & 0 & 0 \\ 0 & 0 & 1 & 1 & 0 & 1 & 0 & 1 & 0 & 0 & 0 & 0 \\ 0 & 0 & 0 & 0 & 1 & 0 & 1 & 1 & 0 & 0 & 0 & 0 \\ 0 & 0 & 0 & 0 & 0 & 1 & 0 & 1 & 0 & 0 & 0 & 0 \\ 0 & 0 & 0 & 0 & 1 & 1 & 1 & 0 & 0 & 0 & 0 & 0 \\ 1 & 1 & 0 & 0 & 0 & 0 & 0 & 0 & 0 & 1 & 1 & 1 \\ 1 & 1 & 0 & 0 & 0 & 0 & 0 & 0 & 1 & 0 & 1 & 1 \\ 1 & 1 & 0 & 0 & 0 & 0 & 0 & 0 & 1 & 1 & 0 & 1 \\ 1 & 1 & 0 & 0 & 0 & 0 & 0 & 0 & 1 & 1 & 1 & 0 \end{bmatrix}.$$

This $(0, 1)$-matrix represents a new graph consisting of the same set of nodes as the network under analysis, but in which two nodes are connected if and only if their intra-cluster communicability is larger than the inter-cluster one. This communicability graph is illustrated in Figure 21.14.

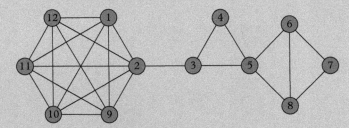

Figure 21.14 *The communicability graph for G_{clus}*

We now find the cliques in the communicability graph, which correspond to the following groups of nodes:

$$C_1 = \{1, 2, 9, 10, 11, 12\}, \quad C_2 = \{3, 4, 5\}, \quad C_3 = \{5, 6, 8\}, \quad C_4 = \{6, 7, 8\}.$$

These cliques correspond to the communities identified by the communicability method and they can be represented by overlapping sets as shown in Figure 21.15.

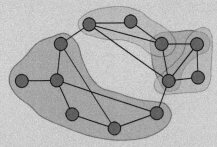

Figure 21.15 *Overlapping communities in G_{clus}*

We want to analyse whether this partition of G_{clus} into four (overlapping) communities is a better representation than those obtained previously which partition the network into two communities, so we calculate the modularity of the new partition. The number of edges and sum of degree in each community are $|E_{C_1}| = 7$, $|E_{C_2}| = |E_{C_3}| = |E_{C_4}| = 3$,

$$\sum_{j \in V_{C_1}} k_j = 16, \quad \sum_{j \in V_{C_2}} k_j = 10, \quad \sum_{j \in V_{C_3}} k_j = 12, \quad \sum_{j \in V_{C_4}} k_j = 10.$$

Substituting these values into (21.1) gives

$$Q = \underbrace{\frac{7}{18} - \left(\frac{16}{36}\right)^2}_{C_1} + \underbrace{\frac{3}{18} - \left(\frac{10}{36}\right)^2}_{C_2} + \underbrace{\frac{3}{18} - \left(\frac{12}{36}\right)^2}_{C_3} + \underbrace{\frac{3}{18} - \left(\frac{10}{36}\right)^2}_{C_4} = 0.426,$$

which is larger than that of $Q = 0.383$ previously found by the partitions introduced by the methods considered earlier: Kernighan–Lin, Girvan–Newman, and the two spectral clustering techniques. Consequently, in this particular case, the partition introduced by communicability is, at least in terms of modularity, better than the bipartition previously considered.

21.7 Anti-communities

In Chapter 18 we showed how to measure the degree of bipartivity in a network. We now show how to find bipartitions in complex networks. For obvious reasons we call these clusters *anti-communities* in the network.

Examples 21.11

(i) In Figure 21.16 we illustrate a protein–protein interaction (PPI) network in which a certain level of bipartivity is expected, since some proteins can act as locks and others as keys in the formation of noncovalent interactions.

(ii) In Figure 21.17 we illustrate a network of employees in a work place, some of whom can be classified as advisers and others as advisees. A degree of bipartivity structure is expected: most (but not all) interactions will between advisors and advisees.

continued

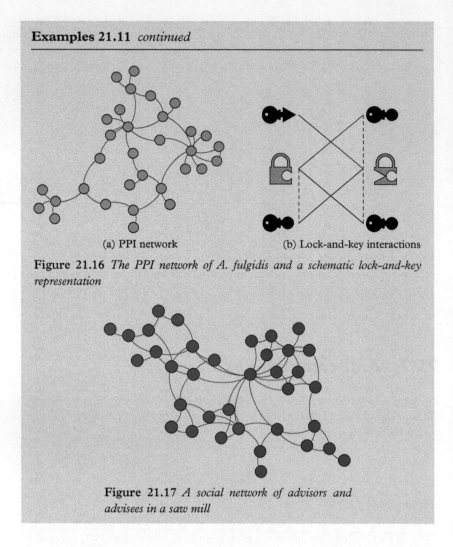

(a) PPI network   (b) Lock-and-key interactions

Figure 21.16 *The PPI network of A. fulgidis and a schematic lock-and-key representation*

Figure 21.17 *A social network of advisors and advisees in a saw mill*

To find such anti-communities, we start by defining an anti-communicability function by

$$\widehat{G}_{pq} = [\exp(-A)]_{pq} = [\cosh(A) - \sinh(A)]_{pq}. \qquad (21.6)$$

Consider a bipartite network and let p and q be two nodes which are in the two different disjoint partitions of the network. Since there are no walks of even length starting at p and ending at q,

$$\widehat{G}_{pq} = [-\sinh(A)]_{pq} < 0. \qquad (21.7)$$

However, if p and q are in the same partition then since there are no walks of odd length connecting them, due to the lack of odd cycles in the bipartite graph,

$$\widehat{G}_{pq} = [\cosh(A)]_{pq} > 0. \tag{21.8}$$

Consequently the sign of $\widehat{G}_{pq}$ determines whether the corresponding pair of nodes are in the same partition or not. That is, $\widehat{G}_{pq} > 0$ if and only if p and q are in the same partition.

Definition 21.4 *Let $C \subseteq V$ be a cluster of nodes in the network. Then C is a quasi-bipartite cluster if and only if $[\cosh(A)]_{pq} > [\sinh(A)]_{pq}$ for all $p, q \in C$.*

We can adapt the procedure we developed for detecting communities based on the communicability function, but this time using anti-communicability replacing G with $\widehat{G}$.

- Form the anti-communicability matrix $\widehat{G} = \exp(-A)$.
- Define

$$\widehat{G}'_{pq} = \begin{cases} 1, & \widehat{G}_{pq} > 0, \\ 0, & \widehat{G}_{pq} \leq 0 \text{ or } p = q. \end{cases}$$

- Let $\widehat{G}'$ represent the adjacency matrix of an anti-communicability graph. Find the cliques in this graph.
- Each clique in the anti-communicability graph is an anti-community in the network.

Example 21.12

Consider the network illustrated in Figure 21.18. This graph has spectral bipartivity $b_S = 0.64$. Then

$$\widehat{G} = \begin{bmatrix} 10.15 & 6.83 & 7.09 & 7.11 & 6.66 & 8.19 & -5.62 & -7.98 & -7.29 & -7.28 & -9.02 & -8.73 \\ 6.83 & 8.03 & 4.17 & 5.87 & 5.43 & 6.72 & -6.31 & -4.32 & -6.56 & -5.65 & -7.28 & -7.04 \\ 7.09 & 4.17 & 8.05 & 5.62 & 5.50 & 6.73 & -5.42 & -6.62 & -4.30 & -6.56 & -7.29 & -7.05 \\ 7.11 & 5.87 & 5.62 & 8.17 & 3.93 & 6.80 & -5.68 & -5.96 & -6.62 & -4.32 & -7.98 & -7.10 \\ 6.66 & 5.43 & 5.50 & 3.93 & 7.37 & 6.36 & -5.39 & -5.68 & -5.42 & -6.31 & -5.62 & -6.77 \\ 8.19 & 6.72 & 6.73 & 6.80 & 6.36 & 9.38 & -6.77 & -7.10 & -7.05 & -7.04 & -8.73 & -7.28 \\ -5.62 & -6.31 & -5.42 & -5.68 & -5.39 & -6.77 & 7.37 & 3.93 & 5.50 & 5.43 & 6.66 & 6.36 \\ -7.98 & -4.32 & -6.62 & -5.96 & -5.68 & -7.10 & 3.93 & 8.17 & 5.62 & 5.87 & 7.11 & 6.80 \\ -7.29 & -6.56 & -4.30 & -6.62 & -5.42 & -7.05 & 5.50 & 5.62 & 8.05 & 4.17 & 7.09 & 6.73 \\ -7.28 & -5.65 & -6.56 & -4.32 & -6.31 & -7.04 & 5.43 & 5.87 & 4.17 & 8.03 & 6.83 & 6.72 \\ -9.02 & -7.28 & -7.29 & -7.98 & -5.62 & -8.73 & 6.66 & 7.11 & 7.09 & 6.83 & 10.15 & 8.19 \\ -8.73 & -7.04 & -7.05 & -7.10 & -6.77 & -7.28 & 6.36 & 6.80 & 6.73 & 6.72 & 8.19 & 9.38 \end{bmatrix},$$

continued

Example 21.12 *continued*

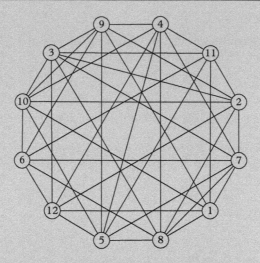

Figure 21.18 *A network with anti-communities*

and

$$\widehat{G}' = \begin{bmatrix} 0 & 1 & 1 & 1 & 1 & 1 & 0 & 0 & 0 & 0 & 0 & 0 \\ 1 & 0 & 1 & 1 & 1 & 1 & 0 & 0 & 0 & 0 & 0 & 0 \\ 1 & 1 & 0 & 1 & 1 & 1 & 0 & 0 & 0 & 0 & 0 & 0 \\ 1 & 1 & 1 & 0 & 1 & 1 & 0 & 0 & 0 & 0 & 0 & 0 \\ 1 & 1 & 1 & 1 & 0 & 1 & 0 & 0 & 0 & 0 & 0 & 0 \\ 1 & 1 & 1 & 1 & 1 & 0 & 0 & 0 & 0 & 0 & 0 & 0 \\ 0 & 0 & 0 & 0 & 0 & 0 & 0 & 1 & 1 & 1 & 1 & 1 \\ 0 & 0 & 0 & 0 & 0 & 0 & 1 & 0 & 1 & 1 & 1 & 1 \\ 0 & 0 & 0 & 0 & 0 & 0 & 1 & 1 & 0 & 1 & 1 & 1 \\ 0 & 0 & 0 & 0 & 0 & 0 & 1 & 1 & 1 & 0 & 1 & 1 \\ 0 & 0 & 0 & 0 & 0 & 0 & 1 & 1 & 1 & 1 & 0 & 1 \\ 0 & 0 & 0 & 0 & 0 & 0 & 1 & 1 & 1 & 1 & 1 & 0 \end{bmatrix}.$$

The anti-communicability graph with adjacency matrix $\widehat{G}'$ is the disconnected network illustrated in Figure 21.19. This anti-communicability graph has only the two cliques

$$C_1 = \{1, 2, 3, 4, 5, 6\}, \quad C_2 = \{7, 8, 9, 10, 11, 12\}.$$

These two cliques are the anti-communities in the network. This can be emphasized by redrawing the network as in Figure 21.20.

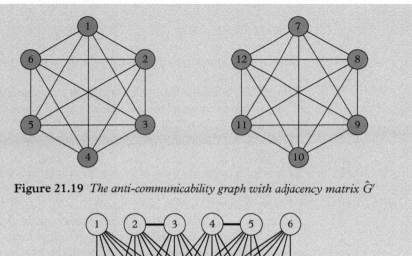

Figure 21.19 *The anti-communicability graph with adjacency matrix* $\hat{G}'$

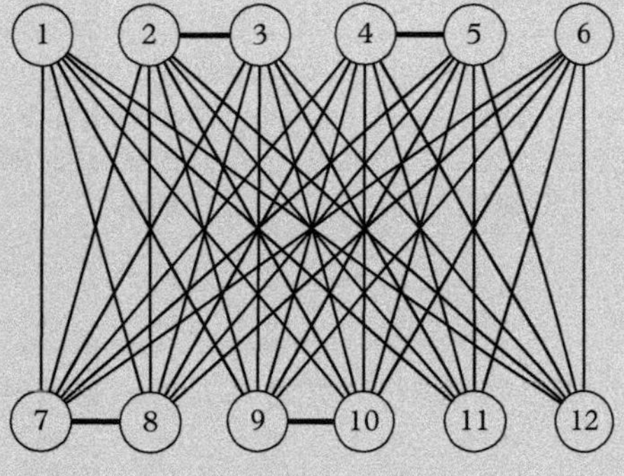

Figure 21.20 *The best partition into anti-communities of the network in Figure 21.18*

Using the approach detailed in this section, we can find the best bipartitions for the PPI and the advisors–advisees networks illustrated in Figure 21.16. The results are shown in Figure 21.21.

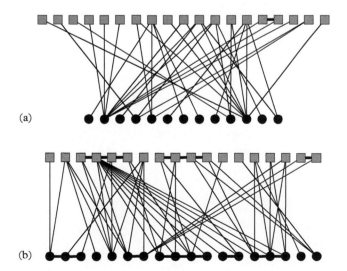

(a)

(b)

Figure 21.21 *Anti-community partitions of (a) the PPI network in Figure 21.16 and (b) the sawmill network in Figure 21.17*

....

FURTHER READING

Estrada, E., Community detection based on network communicability, Chaos 21:016103, 2011.

Estrada, E., Higham, D.J., and Hatano, N., Communicability and multipartite structure in complex networks at negative absolute temperature, Physics Review E 78:026102, 2008.

Fortunato, S., Community detection in graphs, Physics Reports 486:75–174, 2010.

Leskovec, J., Lang, K.J., and Mahoney, M., Empirical comparison of algorithms for network community detection, WWW'10 Proc. 19th Int. Conf. on World Wide Web, ACM, New York, 2010, 631–640.

Ma, X., Gao, L., and Yong, X., Eigenspaces of network reveal the overlapping and hierarchical community structure more precisely, Journal of Statistical Mechanics: Theory and Experiment PO8012, 2010.

Porter, M.A., Onnela, J.-P., and Mucha, P., Communities in networks, Notices of the AMS 56:1,082–1,097, 2009.

Appendix

We include this appendix to add a little detail to certain assumptions made in the text. This detail can be omitted on first reading.

Firstly, note that in Chapter 11 we make the simplifying assumption that for an ER random graph, all eigenvalues apart from the largest can be neglected in understanding certain behaviours as the number of nodes increases. In fact the precise behaviour is more subtle.

For an ER random graph, we know that $\lim_{n \to \infty} \frac{\lambda_1}{n} = p$, and $\lim_{n \to \infty} \frac{\lambda_2}{n^\varepsilon} = 0$ for $\varepsilon > 0.5$. Thus,

$$\lim_{n \to \infty} \left(\frac{\lambda_1}{n} - \frac{\lambda_2}{n^\varepsilon} \right) = p,$$

which means that in the limit $\lambda_1 \geq n^\alpha \lambda_2$, for some $\alpha < 0.5$. The exponent α depends on the density of the network. For networks with very low density α is small, but as soon as the density of the network increases, this exponent approaches 0.5 asymptotically. This means that the spectral gap in an ER network is

$$\lambda_1 - \lambda_2 \geq (n^\alpha - 1)\lambda_2,$$

indicating that $\lambda_1 - \lambda_2$ grows with the density of the network. For instance, for ER networks with density 0.008 the spectral gap is about 3, while for density 0.08 it is about 60.

This asymptotic behaviour does not alter the qualitative results we derive making our simplifying assumption.

Secondly, in Chapter 18 we introduce the quantity

$$\alpha(G) = \alpha = \frac{\sum_{j=1}^{n} \sinh(\lambda_j)}{\sum_{j=1}^{n} \cosh(\lambda_j)},$$

to analyse the effects on bipartivity of adding an edge to a network with approximate bipartition $E = V_1 \cup V_2$ in order to derive the condition $bEE^{even} \geq aEE^{odd}$. This condition indicates that the edge added to G gives a greater contribution to odd than to even closed walks relative to the ratio of these walks in the original graph G. This is always the case when the additional edge links nodes in the same partition of G. That is, when the edge $e = (i,j)$ is such that $i \in V_k$ implies that $j \in V_k$, $k = \{1, 2\}$. If e connects two nodes in different partitions then it is possible that $b < \alpha a$. In this case it is possible that the bipartivity of the network can increase.

Index